Communication Strategies for Engaging Climate Skeptics

Communication Strategies for Engaging Climate Skeptics: Religion and the Environment examines the intersection of climate skepticism and Christianity and proposes strategies for engaging climate skeptics in productive conversations.

Despite the scientifically established threats of climate change, there remains a segment of the American population that is skeptical of the scientific consensus on climate change and the urgent need for action. One of the most important stakeholders and conversants in environmental conversations is the religious community. While existing studies have discussed environmentalism as a factor within the religious community, this book positions religion as an important factor in environmentalism and focuses on how identities play a role in environmental conversation. Rather than thinking of religious skeptics as a single unified group, Emma Frances Bloomfield argues that it is essential to recognize there are different types of skeptics so that we can better tailor our communication strategies to engage with them on issues of the environment and climate change. To do so, this work breaks skeptics down into three main types: "separators," "bargainers," and "harmonizers." The book questions monolithic understandings of climate skepticism and considers how competing narratives such as religion, economics, and politics play a large role in climate communication. Considering recent political moves to remove climate change from official records and withdraw from international environmental agreements, it is imperative now more than ever to offer practical solutions to academics, practitioners, and the public to change the conversation. To address these concerns, this book provides both a theoretical examination of the rhetoric of religious climate skeptics and concrete strategies for engaging the religious community in conversations about the environment.

This book will be of great interest to students, scholars, and practitioners of climate change science, environmental communication, environmental policy, and religion.

Emma Frances Bloomfield is an Assistant Professor in the Department of Communication Studies at the University of Nevada, Las Vegas, USA.

Routledge Advances in Climate Change Research

For more information about this series, please visit: www.routledge.com/
Routledge-Advances-in-Climate-Change-Research/book-series/RACCR

Communication Strategies for Engaging Climate Skeptics

Religion and the Environment

Emma Frances Bloomfield

LONDON AND NEW YORK

from Routledge

First published 2019 by Routledge

2 Park Square, Milton Park, Abingdon, Oxon, OX14 4RN

605 Third Avenue, New York, NY 10017

Routledge is an imprint of the Taylor & Francis Group, an informa business

First issued in paperback 2020

British Library Cataloguing-in-Publication Data
A catalogue record for this book is available from the British Library

Library of Congress Cataloging-in-Publication Data
A catalog record has been requested for this book

ISBN: 978-1-138-58593-5 (hbk)
ISBN: 978-0-367-72962-2 (pbk)

Typeset in Times New Roman
by Wearset Ltd, Boldon, Tyne and Wear

Contents

Acknowledgments

This book stems from my dissertation work at the University of Southern California, so I want to acknowledge all of my peers, colleagues, advisors, and mentors at Annenberg who helped my scholarship come to life. I am especially indebted to my advisor, Randy Lake, for supporting me throughout the conceptualization and writing of this book. From the mountains of Alta, Utah to the castles of Uppsala, Sweden, Randy is a lifelong mentor, colleague, and friend. I am also incredibly thankful to Tom Goodnight, Stephen O'Leary, and Tom Hollihan for their mentorship and guidance while I was at USC and beyond. I would also like to thank the support and comradery of my colleagues at UNLV, who keep my spirits and confidence high. Last, but not least, I'd like to acknowledge the lifelong support of my amazing family, Mom and Dad, Graeme, Trevor, Molly, and Charley, and my loving husband, Paul. I am who I am and my work is what it is because of the support and love I receive from all of those around me. I want to specifically dedicate this book to my Grandma, Pauline Harry, who was an incredible woman who inspired me and everyone she knew to pursue our dreams.

Introduction

The intersection of Christianity and the environment

Artist Adrián Villar Rojas's installation, "The Theater of Disappearance," features a series of tall, narrow sculptures made up of layers of rock, dirt, fabric, household objects, and debris. As one's gaze moves up the sculptures, Villar Rojas's layered towers show the remnants of human influence, such as shoes and technology fading from view. Temporarily housed in the Geffen Contemporary at the Museum of Contemporary Art in Los Angeles from October 2017 to May 2018, these art pieces question the hierarchy of humans controlling the Earth. Over the passage of time, humans themselves will disappear from the geological record despite our current deleterious impact on the environment. The installation portrays humans as but one part of the Earth's history and that the Earth will continue long after we will, although the remnants and scars of our presence might remain. Described as "post-human artwork" (Adrián Villar Rojas: The Theater of Disappearance n.d.: para. 1), Villar Rojas's work represents contemporary tensions over the resiliency of the Earth, the interconnectedness of life, and the relationship between humanity's existence and influence in geological time.

Villar Rojas's installation debuted at a time of polarization in the United States over climate change and proper environmental policies. Embroiled in environmental discourse are questions about the imbalance between human–nature relationships and about how best to negotiate a variety of perspectives about and interpretations of climate science. Some people are highly worried about, others highly skeptical of, and many with feelings in between about what is popularly known as climate change, global warming, and climate disruption (Leiserowitz et al. 2011). There is no doubt or uncertainty in scientific and technical spaces that human actions and behaviors have influenced the Earth to the point of disrupting its natural systems (Ceccarelli 2011; IPCC 2007). Human intervention in the environment has been equated to a "geological force" in that we have littered the world with "human effects and artefacts" and have left traces of human life, not unlike Villar Rojas's sculptures, that have affected the way it functions (Rickards 2015: 283). But, wavering public concern over prioritizing the environment (Pew Research Center 2009) and the reluctance of politicians to take a firm environmental stance foment discussion about what comes next.

A disconnect exists between those that are concerned and alarmed about the degradation of the environment and those that deny humanity's role in climate change. Unfortunately, the vocal presence of skeptics and deniers is loud enough to forestall productive deliberation and to shift conversations to whether climate change is occurring and to what extent, instead of what should be done about it. Those who do trust in the mainstream scientific data may be overwhelmed by the prominence of skeptical voices or might feel like their actions are insignificant to solve the monumental crisis that now faces us. These tensions are particularly prominent in the United States, which houses climate deniers in all levels of government (Call Out Climate Change Deniers n.d.; The Climate Change Deniers in Congress n.d.) and which welcomes industry representatives as powerful players in environmental discussions. Political and public roadblocks to environmental advocacy are seemingly intractable, leaving the balance between humans and the environment uncertain.

Scholars have repeatedly addressed the nature of climate change skepticism and denial in hopes of learning more about why people reject the scientific consensus on the existence and severity of climate change. Skepticism has been contributed to industry strategies of delaying policy action (Ceccarelli 2011), conservative think tanks funding alternative voices (Jacques *et al.* 2008), general ignorance about climate change and its consequences (Bauer *et al.* 2007), journalism practices that keep climate skeptics in the media (Antilla 2005; Kortenkamp and Basten 2015), political bias (Hart and Nisbet 2012), digital discourse (Bloomfield and Tillery 2019; Kirk 2018b), and a variety of other causes. In this book, I add to existing work on climate change skepticism by addressing a facet of skepticism that is rooted in faith.

Specifically, I discuss American Christianity as a potential source of skepticism and a competing narrative to the one that mainstream science tells about the natural world. Within this dynamic, I also explore the ways that Christianity intersects with environmentalism in positive ways. Instead of positing that religion and Christianity are necessarily and consistently anti-science and anti-environmentalism, I provide a detailed analysis of the many ways that religion and environmentalism intersect to influence people's attitudes, beliefs, and behaviors. Sometimes, a religious identity can contribute to skepticism and denial (White Jr. 1967), but in other situations, the intersection of faith and the environment provides opportunities for engagement and progress (Bloomfield forthcoming; Prelli and Winters 2009). Instead of taking a deterministic approach to faith and denouncing its presence in public discourse, I explore the variety of ways that religious rhetoric emerges in environmental communication and how the Christian identity can foster a variety of ecological attitudes.

My approach to climate change skepticism and environmental rhetoric has been influenced by the work of previous scholars who have called for more integration of deliberation and dialogue in academic work. For example, Ceccarelli (2011: 201) argued that scholars should examine "the tools to respond adequately to deceptive arguments about science in public forums" to "increase the discursive choices available" for scientific and environmental exchanges. Instead of examining how scientists communicate about the environment, I offer some practical advice for improving environmental communication based on how the public comes to know and talk about the environment. In this sense, I acknowledge the importance of public participation in environmental issues (Endres 2009) and that we must be both theoretically and practically-minded in environmental communication research. To summarize the imperative for addressing such concerns, I turn to Condit *et al.* (2012: 397) who argued,

> It will likely be important for achieving better human futures if more rhetoricians and members of allied disciplines participate in these efforts to understand science-public interactions, to challenge scientific rhetorics where they are problematic or unjustified, to improve scientific rhetorics where they can be improved, and to build a theoretical structure to support these efforts.

Along these lines, I approached analyzing religious discourse on the environment and engaging dialogue partners through the perspective of rhetorical listening (Ratcliffe 2005), emphasizing dialogue and trust (Goodwin and Dahlstrom 2014), and radical civility (Kirk 2018a). I will elaborate more on these concepts later, but their shared basic tenet is that we should engage people authentically, with the intention of listening and finding common ground. This approach incorporates rhetorical theories that conceptualize rhetoric as a series of interpersonal interactions (Meyer 2012) and turns to both private- and public-level interactions as a means for promoting productive climate conversations.

By exploring religion as a factor in climate change skepticism, this work takes seriously the importance of religious identity in the United States and the variety of ways in which religion functions as an argumentative resource from which people form opinions about the environment and nature, and thus their view of appropriate environmental behaviors. This research is rooted in the power of rhetoric, which I define as the art of purposeful symbol use, both to explain the variety of ways that Christianity influences attitudes toward the environment and to engage people who have a variety of perspectives on environmental issues. In my research, three patterns of integration between Christianity and the environment emerged, which I labeled as three categories, the separators, the bargainers, and the harmonizers. Because I consider climate change skepticism to be a dynamic psychological and rhetorical problem that is rooted in people's values, beliefs, and perspectives, I argue that it is imperative to address one's position on climate change not as a result of a simple lack of information,[1] but as a deeply rooted sign of one's identity. Certain identities seem to align with climate skepticism, such as being a small business owner, a Republican, or a Christian (Gauchat 2012), providing support for the role that identity factors play in the climate change controversy. I propose that my categories provide an identity-focused typology of how people integrate Christianity and the environment to guide their understanding of and response to climate change.

Typology of separators, bargainers, and harmonizers

Throughout the course of my research, I developed three categories to represent the types of responses that occur when Christians are met with climate change information. These categories are not meant to be exclusive or exhaustive but are representative of the prominent rhetorical resources that people and groups turn to in order to make sense of the environment in light of their faith. When incorporating climate change information into their worldview, I found that people turn toward arguments that legitimize the rejection, modification, or acceptance of that information. Arguments of rejection seek to undermine the authority and validity of the new information and retain the integrity of previous perspectives. Arguments of modification seek to adapt and negotiate the new information to more closely align with previous perspectives, potentially damaging the intended meaning or content of the new information. Arguments of acceptance seek to validate the new information by integrating it with previous perspectives, sometimes under a new worldview. I argue that religious adherents' orientating framework toward the environment and its authorities (e.g., climate scientists, media, and the government) largely guide their response to information about climate change. The three types of responses I outline above are labeled the separators, the bargainers, and the harmonizers.

The typology of separators, bargainers, and harmonizers seeks to capture groups and actors who primarily or largely respond to climate change information with rejection, modification, and acceptance, respectively. Those who primarily use strategies of rejection are called separators because their motivation behind rejection is rooted in a perceived conflict between environmentalism and Christian values that separates them into opposing forces. Those who primarily use strategies of modification are called bargainers because their motivation behind modification is to borrow aspects and features of environmental arguments and environmental science to blend with their religious values. These first two categories represent our traditional notion of climate skeptics[2]: those who do not fully (or at all) accept the consensus of climate scientists and are resistant to pro-environmental policies.

While I would not label harmonizers climate skeptics, they are included as a group in this study for two reasons. First, the harmonizers expand the scope of this book to encapsulate a variety of ways that Christians think and act toward the environment. My analysis of the harmonizers serves as a foil to the previous two categories to explore how religion and the environment can be integrated for productive deliberation and activism. The harmonizers represent an interpretation and performance of the Christian identity that is not deterministic of negative attitudes toward climate change. Second, the harmonizers are an exemplar of how belief in climate change, even when supported by a religious mandate, does not always lead to active environmental advocacy and action. There is oftentimes a gap between awareness of the problem of climate change and acting upon it, which is representative of many harmonizers who share a belief in the importance of the environment but may not know how to engage those beliefs practically. The strategies for addressing harmonizers, therefore, focus on encouraging action on already agreed upon principles instead of spurring a reconsideration of their interpretation of the Bible with regards to the environment.

Now that I have provided more details about my typology, it is important to address how acts of classification are "necessarily selective, arbitrary, and consequential" (Ott and Dickinson 2013: 11).[3] I do not intend for this typology to become "objective, generalizable findings" (Jensen 2016: 11) and do not make such assertions. I agree with scholars who have come before me that there is little utility in creating typologies that try to fully encapsulate issues as complicated as scientific controversy (Ceccarelli 2005; Goodnight 2005). I do not believe, however, that the call for specificity and attention to unique situations and circumstances precludes the usefulness of exploring rhetorical patterns that emerge in discourse about controversial topics. These labels, therefore, are not exhaustive exercises in identifying people and groups, but are useful heuristics for navigating the multifaceted beliefs of climate skeptics and how religious values and interpretations influence those various beliefs. If there is no one solution to climate change and if there is not enough time or resources to craft individual messages, then a meso-level approach that makes use of audience segmentation and targeted communication seems to be a fruitful compromise.[4] The groups and individuals I have studied in writing this book serve as examples of how these beliefs manifest themselves in discourse so that we might learn potential strategies for engagement.

To address theoretical concerns about overgeneralization, I have immersed myself in the communities that I am studying and allowed frequent motifs and metaphors to guide my characterizations. In this book, I invoke conceptions of rhetoric as a "dynamic" and "constitutive" process that starts with "acts" instead of actors (Jenkins and Cisneros 2013: 87). From this approach, I start with observing and interrogating the rhetorical activities of certain groups and then form conclusions from those rhetorical features. This approach privileges "contextual evaluations" of rhetoric over a priori judgments or classifications that predetermine what one hopes to find within rhetorical activity (Jenkins and Cisneros 2013: 87). The categories I have outlined are based on exploration of the existing discourse and the identification of salient patterns and trends within the "larger rhetorical ecology" of climate change skepticism (Jensen 2016: 11). I believe that categorization and classification can be important rhetorical tools, especially when they invite modification, fluidity, critical evaluation, and argument (Brockriede 1974) in the pursuit of further knowledge and understanding. Burke (1966: 50) argued that we must embrace and be reflective about the limits of our vocabulary, "since we can't say anything without the use of terms."

Because I wish to intervene in the topic of environmentalism and religion, I aim to use terms carefully and prudently to explain the patterned differences in religious responses to climate change that I found in my research. While I intended simply to learn more about climate skepticism from a religious perspective, I ended up discerning terms and patterns that distinguish certain perspectives and their argumentative resources from one another. I justify these categorization decisions based on the groups' guiding narratives and on rhetorical theory and dramatism. Burke (1966: 49) specifically mentions two types of terms people use, those "that put things together and terms that take things apart." It is my hope that my categorizations do both. I hope to put together patterns of responses to create engagement and to take apart our preconceived ideas about the relationship between Christianity and the environment. In attending to nuances and subtleties within the monolithic term "climate skepticism," we may uncover new rhetorical in-roads and opportunities for engagement. Some scholars have seemingly abandoned Christianity as an outlet for environmental activism, lumping Christians into a homogenous group that is anti-environment (Clements *et al.* 2014; Konisky 2018), perhaps in part due to their apocalyptic beliefs (Barker and Bearce 2013). I hope that this book expands the limits of our current conceptions of our climate crisis and the role that religion plays in it, so that we may not be overly reductive in our view of what skepticism means and how Christian beliefs influence ecocultural identity formation (Bloomfield forthcoming).

I use the terms separators, bargainers, and harmonizers to talk about specific groups and people that make predominant use of certain rhetorical features. I do not mean to imply, however, that these features are exclusively used by each category. People may use multiple rhetorical resources across these categories simultaneously, over time, or never. These categorizations serve to highlight some of the rhetorical resources available to religious adherents and to explore patterns in their usage that constitute sense-making narratives for those who use them. These labels are ways of helping us organize and navigate the analysis of a complicated intersection of religious beliefs and environmental attitudes. Furthermore, I use these groupings as theoretical foundations for the application of targeted strategies to engage religious adherents in environmental advocacy or at least to reimagine how their faith might overlap with environmentalism.

I advocate that by knowing "why people do as they do, we feel that we know what to expect of them and of ourselves, and we shape our decisions and judgments and policies to take such expectancies into account" (Burke 1984: 18). People that view the world differently and use distinct words to describe their reality should not be expected to react to new information in the same ways. Based on the type of rhetorical resources being used, I provide tailored argument strategies to start conversations and maintain engagement despite discussing a topic as contentious as climate change. These strategies are practical suggestions for how people can engage skeptics on a personal and societal level. The term "strategies" can imply intentionality or manipulation, but I use the term to describe the rhetorical moves and "maneuvers" that people use to make sense of and explain their views on their environment (Burke 1974: 298).

My attempt to categorize climate skeptics is not itself novel. Acknowledging that categorizing climate skepticism is "fraught with difficulties," Matthews (2015) offers a typology based on the severity of one's rejection of climate change. From being "lukewarm," to "moderate," and then "strong," he describes the strength of one's denial as being discrete categories (Matthews 2015: 157, 158). In order to place skeptics in these boxes, Matthews (2015) discusses argument claims that each group makes, delineating moderate skeptics as believing the consequences of climate change have been overblown, while strong skeptics believe there has been scientific fraud. My issue with this typology is it seems reasonable that one skeptic could hold both beliefs (e.g., that a fabrication of scientific data has led to exaggerated claims of severity) and thus would seem to fit in multiple categories. Creating categories based on a scale also seems to overlook the variety of degrees in which strength of denial may take shape and to propose that the strength of denial is fixed. In other words, Matthews's (2015) typology seems to make discrete categories out of a continuous relationship.

By basing categories solely on the strength of skepticism, I note that Matthews's (2015) taxonomy ignores the possibility of using multiple argument claims to craft one's perspective on climate change and reduces one's strength of denial to the arguments they use. Similar typologies (e.g., Hobson and Niemeyer 2013; Poortinga *et al.* 2011) also run into this issue of plasticity by creating taxonomies based on the type of doubt exhibited and what about climate science is cause for doubt.[5] This is not to say that these typologies are without value. There may be instances where labeling skeptics based on the strength of their denial helps to distinguish them from other kinds of skeptics, such as Kirk's (2018b) work with Yale Climate Connections in locating those open to engaging in discussion and deliberation.

I wish to build on these typologies by not focusing on strength of denial but by the motivations and worldviews that drive the negotiation of Christianity and the environment. This typology is unique in that it offers explanation as well as description based on underlying ideologies and values. Instead of focusing on specific arguments that are made, this approach acknowledges that certain argument claims may be used frequently across skeptical communities, and thus might not create meaningful categorization. It is certainly important to pay attention to the types of arguments that skeptics make, which makes up a portion of the current inquiry. But, I argue it is more important to interrogate the rhetorical justification *behind* the argumentative resources, because they provide deeper insights and more predictive power. Second, my typology centers on the relationship between religion and faith, recognizing a specific worldview that influences people's attitudes toward the environment. This more targeted focus enables more nuanced understanding of the specific rhetorical features of these religious actors and the argumentative resources to which they turn for environmental attitudes and actions. I do not wish to eliminate or substitute previous typologies completely, but to add mine to the variety of heuristics at our collective disposal for understanding and interrogating people's attitudes toward the environment.

In addition to research elucidating different types of skeptics, scholars have also proposed and tested strategies for engaging climate skeptics.[6] This book adds to existing strategy literature by synthesizing these strategies, proposing new strategies, and categorizing when and to whom these strategies may be most effective. The conversation strategies proposed in the following chapters to engage climate skeptics and Christians in environmental topics are rooted in rhetorical theory and my personal engagement with skeptics and Creation Care members in interviews and surveys. In addition to providing a new typology and concrete strategies, this book contributes to rhetorical theory and environmental communication. As a methodological contribution, I combine argument strategies and dramatism to explore the relationship between the stories we tell and the inventional resources they provide. As a theoretical contribution, I test the utility of rhetorical listening when applied to a complicated topic such as climate change and explore the patterned rhetorical expressions in the climate change controversy as a preliminary step to uncovering patterns and themes that may apply across scientific controversies. In what follows, I highlight these theoretical contributions in more detail by introducing the rhetorical terms used to analyze religious environmental discourse in the upcoming chapters.

Dramatism, argumentation, and rhetorical listening

I have already briefly introduced rhetorician Kenneth Burke, whose theories of language, rhetoric, and worldviews guide this book. Burke's work centers on the symbolic choices people make, the vocabulary available to them, and the resulting blindness toward other perspectives based on our symbolic choices. Burke (1966: 3) defined humans as the "symbol-using animal," connecting humans to our animalistic nature but also recognizing our differences because of our capacity for symbolic reasoning. For Burke, a seemingly innocuous choice of naming or labeling both produces and reflects a specific attitude or behavior. He argues, "Names shape our relations with our fellows. They prepare us *for* some functions and *against* others, *for* or *against* the persons representing these functions" (Burke 1969: 4, emphasis in original). In other words "a situation may be so described that one particular kind of act or attitude is implicit in it" (Burke 1951: 202). To call someone a conversation partner or a friend is a label that connotes respect, understanding, and familiarity. To call someone an enemy, a foe, or an even harsher name is a label that connotes disrespect, disengagement, and separation. For Burke (1969: 57), such labels are equivalents to an "attack," because he viewed the act of naming as a form of symbolic action. Our words have material effects, which change the way we view a situation and thus respond to it.[7]

   Burke further argued that our selection of words limits us in the options we are able to see. In our attempts to define and make sense of reality, he noted that we "develop vocabularies that are *selections* of reality" (Burke 1969: 59, emphasis in original). Each selection that we make of one word (friend) over another (foe) is a "*deflection*" of that alternate reality (Burke 1969: 59, emphasis in original). The vocabularies we use create "hortatory" imperatives, which Burke (1966: 44) defines as providing us with proscriptions for what we should and should not do. These choices come together to create a coherent perspective. In other words, the selection of certain terms invites consistency among other choices, so people are not faced with contradictions. These overarching lenses through which we see are called "terministic screens," emphasizing the power our symbolic naming practices have (Burke 1966: 45). A coherent terministic screen contains patterned labels or "forms" for which to describe things, people, and situations, consequently creating an "appetite" or expectation for how to make sense of and deal with them to the form's "satisfaction" (Burke 1968: 31). In discussing terministic screens and scientific discourse, Prelli (1989: 89–90) argued that "once we are induced to accept particular terministic screens, we gain entry to the orientation invoked by those terms. We will treat those terministic screens as unquestioned presuppositions, if only provisionally." Thus, examining the words that people use in the climate change controversy to identify themselves and others provides insight into their worldview, values, and attitudes, which shape appropriate responses.

In this book, I found that the terministic screens of the three categories frequently manifested as metaphors, which are artistic comparisons that transfer characteristics from one domain of experience to another. Metaphors create mental shortcuts and connections between ideas that shape how we view them. On the physical level, metaphors activate neural pathways that create instant, emotional "circuits" between words and ideas (Lakoff 2010: 71). The repetitive application of those metaphors create stronger pathways that transform what was once a "nonliteral comparison" into "normal language" (Foss 2017: 285; Lakoff 2010: 72). Metaphors are made up of two parts, the vehicle and the tenor (Foss 2017). We consider the tenor the topic we are talking about and the vehicle the "lens" through which we understand that topic (Foss 2017: 285). Like a terministic screen, a metaphor can create a "filter" which ascribes certain characteristics to a topic, thus influencing our perceptions of that topic (Foss 2017: 286). Lakoff and Johnson (1980: 454) place the utmost importance on metaphor, arguing that our very "conceptual system in terms of which we both think and act, is fundamentally metaphorical." Metaphors are indicators of people's worldviews, because the language we use reflects the conceptual system that "defin[es] our everyday realities" (Lakoff and Johnson 1980: 454).

For example, a common vehicle through which climate skeptics understand climate scientists is that of prophecy, fortune-telling, and prediction (Bloomfield and Lake 2015: 387). Calling Al Gore a "green prophet of doom" (Musser 2013: para. 4) and making the cover of one's book about climate change a crystal ball (Inhofe 2012) are the verbal and visual remnants of a metaphorical framework that associates science with mysticism, fraud, and alarmism. Metaphorical frames are thus important to environmental discourse, because they help us understand how certain perspectives and information can be ignored and deflected (Lakoff 2010: 73). Lakoff (2010) specifically addressed how Christian frames have been activated to support climate denial, something that will be further explicated in the following chapter about the separators. If we can rewire these connections, offer more convincing ones, and shift the conversation to avoid activating hostile frames, we may begin to have an open, productive conversation about the environment.

An understanding of people's vocabulary, their go-to filters, and guiding metaphors crafts a picture of a person's attitudes, beliefs, and behaviors. The terministic screen or frame through which people view situations will influence how they respond to them. If people make sense of climate change and the environment by seeing the situation as a war between good and evil (i.e., the separators), their vocabulary associates compromise with a betrayal of one's side and thus makes listening to alternative vocabularies incredibly difficult. The constellation of metaphors and screens also provides us with an insight into the argumentative resources likely to be used as a part of the individual's worldview. If people have a particular way of seeing the world, they are likely to make arguments to support and legitimize those worldviews.

Contemporary rhetorical theorists and environmental communication scholars still make use of Aristotle's original conceptions of the persuasive resources available to speakers. Among Aristotle's original contributions to argumentation are the emphasis on enthymemes and rhetorical *topoi* or commonplaces. Enthymemes are a type of argument that relies on a series of premises being given that an audience can adhere to in sequence. After following the sequence, the audience fills in their own perspectives and understandings in order reach the speaker's conclusion. These premises are directly related to people's underlying worldview and frames, because the premises represent statements that the audience adheres to and agrees with.

For example, when speaking to a Christian audience, a speaker can assume the audience's agreement that the Bible is important, valuable, and should be followed. But, they may not know how that person interprets the Bible or which parts are valued over others. In making a case for their conclusion, therefore, they can leave the premise of the Bible's value "unspoken" for the audience to fill in their particular values in the gap. The speaker might talk about charity and giving to the poor as an important theme in the Bible, frequently mentioned and highlighted, and then conclude that good Christians should give to a particular charity. An argument for the value of following Biblical teachings is unnecessary to complete the argument (which fleshed out in its entirety is called a syllogism) and is thus left unsaid. Scholars argue that enthymemes are powerful ways to involve the audience in its own convincing, inviting their direct participation in forming the argument (Bloomfield and Sangalang 2014; Waddell 1990). Because of their widespread use and standard practice in argument theory, enthymemes have been referred to as the centerpiece "of all reasoned discourse" (McBurney 1936: 50). Enthymemes have previously been studied in environmental topics through oft-repeated topics, or rhetorical commonplaces, that "collapse or abbreviate" fully developed arguments (Tillery 2018: 17).

As an extension of examining enthymemes as a form that arguments can take, it is also important to map out the entire structure of arguments to locate opportunities for intervention and refutation. Argument maps aid in isolating how terministic screens and frames influence the selection of argumentative resources. One such tool for mapping arguments is the Toulmin model. The Toulmin model is useful because it provides a systematic way to critique arguments and contributes to proper classification of argument components (Brockriede and Ehninger 1960), both of which are concerns of this book. The model contains three primary parts: claim, grounds, and warrant.

A claim, for example, might be that we should care about the environment. The harmonizers turn to the Bible to find evidence for such a claim, which the model calls the grounds[8] of the argument. The grounds that support the claim of caring for the environment might be that the Bible says that we should and could include particular verses, such as Genesis 2:15[9] as evidence. The third component is the warrant, which is described as the logical "movement" from the data to the claim (Brockriede and Ehninger 1960: 44). To complete this example following the Toulmin model, the movement between verses in the Bible and the claim that we should do something about the environment rests on the logical assumption that the Bible is something to be valued and is a proscription for action. Similar to the premises in our enthymeme example, the warrant of an argument is often an unspoken "mental leap" that "certifies" and validates that the grounds support the claim (Brockriede and Ehninger 1960: 45). A failure to follow the warrant disrupts the grounds–claim relationship and leaves the argument weak and susceptible to counter-arguments.

By mapping the Toulmin models within discourse, we can isolate points of disagreement. Using the example above, we can characterize a harmonizer argument as follows:

Harmonizer argument

Claim: We should care about the environment.
Grounds: Because the Bible contains verses about the importance of environmental protection.
Warrant: The Bible is a valuable source of reasoning and a proscription for action.

Separators reach a different conclusion about the environment, but are authorized by the same warrant:

Separator argument

Claim: We should not prioritize the environment over human needs and wellbeing.
Grounds: Because the Bible contains verses that legitimize human life as more important than nature or animal life.
Warrant: The Bible is a valuable source of reasoning and a prescription for action.

The primary difference between the separators' and harmonizers' arguments is where they locate the grounds for their arguments, while they would share the same warrant. Previous scholars have mapped the overlap between warrants and worldviews by arguing that underlying ideologies legitimize and provide support for the argument's warrant (Bloomfield and Tscholl 2018).

These three components are integral to my later analysis comparing the argument forms afforded by the groups' guiding terministic screens. They also provide opportunities to see overlap and commonalities between the disparate groups, while also recognizing their instructive differences. For example, identifying a skeptic's or interlocuter's grounds provides insight into where they turn for argumentative resources and potential ways to substitute those grounds with competing or complementary grounds. The Toulmin model thus helps us analyze where and how people construct environmental arguments.

Discussing argument and rhetorical concepts prepares us for their application to the words and thoughts of the case study groups and individuals in future chapters. How the three categories differently frame the environment and environmental science shapes their discursive responses to climate change information. My focus on screens, metaphors, and the Toulmin model serves as foundations for the evaluation of potential successful and unsuccessful strategies in addressing these groups as well as creating a holistic look at how rhetorical, argumentative, and social components of identity intersect and manifest in controversy discourse. To complement a Burkean dramatism and argument approach, I also employ theories of rhetoric that emphasize the importance of interpersonal interactions and trust-giving. O'Leary (1993: 399) argued that Burke's approach to appetites and satisfaction are markers of an "audience-centered definition" of argumentation, whereby communal meaning and understanding is developed "in the process of interaction between rhetoric and audience" instead of through overt or direct persuasion. This orientation to argumentation echoes Chaim Perelman's "transcendent" approach to argumentation where the audience is persuaded through a "meeting of minds" with the rhetor (Long 1983: 107, 115).

Focusing on the audience, treating their attitudes as the communicative remnants of their terministic screens and engaging them as dialogue partners (instead of foes or dupes to be persuaded) are integral for achieving deeper understanding of their attitudes toward the environment. In respecting people's terministic screens and rhetorical choices, we make a move against persuasion and toward identification. Burke (1969: 43) argued that identification is when people find common ground or shared values from which to connect and to foster "cooperation." In viewing dialogue as identification instead of persuasion, we ourselves make a rhetorical choice to approach conversations differently, with different goals, and different attitudes. Within this perspective, I used the strategies of rhetorical listening to converse with my dialogue partners, who I mostly found through online discussion boards. Rhetorical listening, as defined by Ratcliffe (1999: 206), proceeds in three steps: "first acknowledging the existence of these discourses; second, listening for the (un)conscious presences, absences, unknowns; and third, consciously integrating this information into our worldviews and decision-making." To employ step one, I approached conversations with skeptics as not discourses to be eliminated, but to be understood. To employ step two, I sought to ask questions beyond the topic of climate change and explore deeper values, priorities, and understandings. To employ step three, I sought common ground and points of similarity with those underlying discourses to move conversations toward cooperation, mutual understanding, and identification.

In addition to engaging ideas of rhetorical listening, I also kept in mind Goodwin and Dahlstrom's (2014) strategies for promoting trust in climate change conversations. Directly aligned with rhetorical listening is their suggestion to "empower your audience," whereby the conversation is not one of one-way instruction, but of mutual knowledge acquisition (Goodwin and Dahlstrom 2014: 156). This strategy produces a frame of mind where we view our dialogue partners as having something to contribute, which also opens them up to have their opinions assessed and held responsible as part of climate conversations. I also frequently used Goodwin and Dahlstrom's (2014: 156) suggestion to "start small," build from low stakes interactions, and separate "issues of policy" from "issues of science" to find areas of commonality and trust. Goodwin and Dahlstrom (2014: 156) note that these strategies are only the "first step in what must be an on-going effort to communicate climate science," but it is an important first step in that conversations may not continue or spark future interactions if trust is lost or damaged.

This chapter has outlined the presence of climate skepticism as a problem facing climate change and environmental communication, justified the need for a new categorization to understand the religious facet of this population, and introduced important rhetorical tools that will be developed and applied in future chapters. What has yet to be discussed is the context from which this research has arisen and why I have chosen to focus on the relationship between environmentalism and Christianity. To explicate that context, I will give a brief overview of how American society has understood "the environment" as separate from its citizens. Of particular importance in this brief overview is the multitude of stories about the environment that live in the American consciousness influenced by and influencing Christianity.

American Christianity and environmentalism

Recent years have not been kind to the environment. Climate skeptics make up nearly a third of the US population (Pew Research Center 2013), there has been an influx of anti-environmentalist politicians operating important environmental posts (Mufson 2017), and some portions of the public have vacillating trust in science as a reliable source of information (Gauchat 2012). The repeated inclusion of contrarian voices to the established consensus on climate change[10] appears to be influencing the public's perceptions on scientific consensus and appropriate actions to take toward the environment (Brüggemann and Engesser 2017). When the Kyoto Protocol was reworked into the Paris Climate Accord, the US initially joined it under Barack Obama. In 2017, however, Donald Trump withdrew from the agreement, making it, at that time, the only country not to have signed it (Friedman 2017). In the wake of Trump's election, the Environmental Protection Agency (EPA)'s website underwent changes, with some topical websites going offline and terms such as "climate change" disappearing from website titles and content (Miller 2017). Current environmental communication seems to be trapped in a perpetual state of addressing the manufactured controversy (Ceccarelli 2011) of the debate over climate change, unable to discuss appropriate actions and environmental policies. Some people attribute the growth of the Internet and social media as a means by which climate denial grows and spreads, but it is hard to isolate a single factor as responsible for a lack of trust in scientific authorities and climate science. At the very least, we can argue that such online communities legitimize the skeptical identity and provide a platform for skeptics to create inclusive, supportive environments (Bloomfield and Tillery 2019).

Many have argued that public discourse and policy implications stem from the unique role of the environment in America's mythos. After Europeans landed here and deemed it uncivilized, it became an obligation of the American identity to explore the land from coast to coast. Manifest Destiny was part of America's "civil religion," in that it constituted a sacred duty to "redeem the continent" and stretch America's boundaries (Coles 2002: 404). America's expansion has become "the most enduring and characteristic American myth," whereby the cowboy, who tamed the wilderness becomes America's "archetypal hero" (Rushing 1983: 15, 16). The metaphor of the frontier and exploration has been a mainstay of American public discourse (Ceccarelli 2013). People like Theodore Roosevelt (Dorsey 1995; Tillery 2018) became iconic environmental figures, primarily because of their marriage of environmental protection with a rugged, individualistic identity. We build monuments (McGeough *et al.* 2015) and museums (Dickinson *et al.* 2005) to our colonial heroes and try to "absolve" ourselves of colonial guilt (Dickinson *et al.* 2006: 28). Landmarks, such as our National Parks, are preserved in part because they conjure up feelings of the "sublime," becoming sacred, protected parts of the American landscape (DeLuca and Demo 2000: 246). The American relationship to the environment is full of tensions between conquering, exploration, preservation, and exploitation. While America's identity comes in part from dominating the wilderness, there remains a reverence for its beauty and a call to protect what remains.

Embedded in this mythos is the notion that humans are separate from nature and the environment, a hierarchical relationship that, in part, enables the exploitation of the environment to be as less important than humanity (Milstein and Dickinson 2012). Nature and humanity, however, are well-established in environmental communication literature as part of the same ecological network, separated by words and labels but not substance (Milstein 2011). Conley and Mullen (2008: 195), for example, argue that a "nature–culture binary" is a rhetorical tool that helps people negotiate their place in the world around them, despite ultimately being a fabrication and "fiction." The perpetual notion of nature as separate from humanity, and even the ways that we often express this relationship through language,[11] reinforce the idea that humans are one thing and the environment is a separate thing altogether. In the case of American, Western mythos, the environment becomes a thing to conquer and control; the wilderness is something wild, something *not* human.[12] The tensions between humanity, nature, culture, the environment, non-human animals, and other ecological factors do not vanish or become clearer when one takes into consideration the added element of Christianity. As the focus of this book, it is integral to interrogate the added complications that arise when religious identity is added to the mixture of what constitutes environmental attitudes and the potential for ecologically "restorative discourse" (Milstein 2012: 162).

Throughout this book, I conceptualize religion as a "sacralization of identity" (Mol 1977: 1) or the process of making sacred one's sense of self. Something that is sacred is, by definition, not something easily violated or changed, but is exalted and integral to one's actions and behaviors. If a terministic screen, metaphor, or framework becomes sacred, it takes on a new level of importance, consecration, and obligation to follow. Religion may be the ultimate terministic screen as it prescribes hortatory (or obligatory) rules for everyday life. Burke (1984) called notions of the sacred one's sense of "piety," or the established order that one tries to follow. New information that appears to violate or question that piety may be viewed as sinful, dangerous, or immoral. Mol (1977: 6) argued that sacralization is an act that "protects identity, a system of meaning, or a definition of reality, and modifies, obstructs, or (if necessary) legitimates change." For Mol, sacralization is both a defensive maneuver and a productive, performative act. To reject the science of climate change, to modify it to fit one's faith, or to accept and integrate it are acts that simultaneously construct an identity and protect that identity. In taking seriously the personal and social implications of religious identity, I conceptualize religious adherents' positions on climate change as outcomes of a complex negotiation of competing values, beliefs, and identities, where the resulting beliefs and behaviors stem from guiding frameworks and how people have come to learn about the world around them.

For some Christians, mainstream scientific conclusions are connected. Under this perspective, a belief in climate change may also open oneself up to belief in evolution or atheism (McCammack 2007). These slippery slope arguments are frequent markers of anti-science discourse (Bloomfield 2017), but are not representative or deterministic of all ways that religious people interact with mainstream science. Evans (2018) makes a convincing argument that religious people are not universally anti-science, but can be highly concerned about certain iterations of science and their implications for their values and priorities. For example, McCammack (2007) argued that skepticism toward climate change from evangelical communities can be attributed, in part, to the belief that accepting environmental conclusions of mainstream science may affect other tenets of their faith. Thus, science and religion are sometimes considered competing epistemologies, or ways of knowing, which can spark conflict when the two sources of information are perceived as contradictory. This framework, however, often exaggerates the divide between the two, as groups traditionally labeled as anti-science, such as creationists, directly appeal to science as grounds for their arguments (Bloomfield 2017; Hoerl and Kelly 2012).

While adherence to a religion is not indicative of being anti-science, there is evidence that belief in Christianity and regular church attendance is negatively associated with pro-environmental beliefs (Gauchat 2012). Lynn White, Jr. (1967: 1205) argued that many Christian values are incompatible with ecology, noting, "What people do about their ecology depends on what they think about themselves in relation to things around them" which is "conditioned by beliefs about our nature and destiny – that is, by religion." The balance between humans and nature, or the sustainability of "humanature" (Milstein *et al.* 2011: 488), is thus directly linked to one's faith. In offering a framework that believes "no item in the physical creation had any purpose save to serve man's purposes," Christianity has been accused of being "the most anthropocentric religion the world has seen" (White Jr. 1967: 1205). Environment scholars Tema Milstein and Elizabeth Dickinson (2012: 511) identified anthropocentric and ecocentric, "nature-centered," discourses as part of a "dialectical pull" between competing interests. Within an anthropocentric perspective, caring for the environment may be viewed as a betrayal of caring for and valuing human life. This distinction reifies the idea that nature and humans are separate, which is also present in the Bible's description of humans having "dominance" (White Jr. 1967: 1205) or "dominion" (Genesis 1:26, King James Version) over all other life.[13] These verses are often viewed as in tension with other verses and parables that provide alternative perspectives on the human relationship to the environment.

Because Christians use the Bible as an argumentative resource, their mobilization of the Bible both for and against pro-environmental action comes down to the rhetorical frameworks that guide their interpretations of the text. These frameworks, or terministic screens, can be influenced by many factors, such as additional identities or ideologies, socialization through friends, family, and the community, or a predilection for skepticism and distrust of authority. Many of the people I interviewed in this book pointed to family, peer networks, their education, personal research, media, and other sources in addition to their faith as playing a role in their environmental attitudes. Many interviewees pointed to politics as influencing in their beliefs, while often simultaneously lamenting that the environment had become embroiled in it. The various interpretations of faith, or hermeneutics, lead religious adherents to consume, process, and respond to climate science in disparate ways.

In the United States, adherence to Christianity is linked to one's politics, where voting decisions may be influenced, in part, by social issues such as gay marriage and abortion rights (Domke and Coe 2010). Many Christian denominations lean conservative (Lipka 2016), and elections from 2000–2016 show that Republicans have continued, strong support from white religious groups, especially evangelicals and Protestants (Smith and Martinez 2016). Political scholar Green (2007) argued that the deterministic relationship between politics and religion is accentuated in evangelical populations. Making up 26 percent of the voting public (Smith and Martinez 2016), evangelicals "identify strongly with their religious affiliation" and believe "their faith is relevant to their politics" (Green 2007: 28). Domke and Coe (2010: 12) further argued that America's "religious heritage ... is increasingly used for partisan purposes," implying that one's faith can be activated and thereby connected to policy agendas. In the constellation of one's identity, core beliefs such as religion and politics can become closely intertwined. In this sense, conservative politics influences the relationship between Christianity and the environment. Considering that conservatism and church attendance are both correlated with a skepticism of science (Gauchat 2012), the story of climate change denial in the United States lies at the intersection of religion, politics, economics, science, and the environment.

It is important to remember when studying these topics that there are no strict causal relationships one can make between religious identity and environmental attitudes, which makes exploring this intersection fruitful, but also complicated. The climate change controversy may not always seem to directly invoke or incorporate religious elements, but religious identity can be an important factor in many people's climate change beliefs. For example, Ross (2013) found that 16 percent of people he interviewed about their environmental beliefs brought up their faith without prompting, indicating that some interviewees saw a direct connection between their faith and the environment. Previous studies have also suggested that apocalyptic beliefs, such as Christian notions of the Second Coming of Jesus, negatively predict pro-environmental attitudes (Barker and Bearce 2013). Conversely, other Christians use their faith as justification and motivation for environmental protection (Bloomfield forthcoming; Prelli and Winters 2009). Untangling the web of connections between faith and the environment entails attending to how religious identity is mobilized and what attitudes are subsequently fostered toward mainstream climate change conclusions.

To understand the turbulent constellation of these factors as they contribute to climate denial in the United States, the following chapters identify key groups and actors that make use of the three strategies of separating, bargaining, and harmonizing religion and environmental science. The featured separator is the Cornwall Alliance (CA), a religious environmental group started in 2005. The featured bargainer is the Acton Institute (AI), a policy think tank started in 1990. The featured harmonizer is the Evangelical Environmental Network (EEN), a Creation Care group started in 1993. I focus on an exemplar group of each type to provide a deep, close reading of their primary texts and to map how their discourse unfolds as an identity performance and explanatory narrative. My artifacts primarily consist of the website content and publications of these groups, online forums for members, and publicly available media content about them. All of these groups have highly curated web pages with historical information, group justification, partner groups, and some have shops where books and other media can be purchased. These websites constitute the official discourse of environmental groups, but it is also important to look at how these discourses circulate and become part of the public conversation (Bloomfield forthcoming; Milstein *et al.* 2011). To this end, I supplement my analysis of all three categories with interviews and surveys with members of the groups and individuals who exhibit rhetorical features of these categories. These participants were recruited through social media and invited to engage in conversation via text, phone, and video calls.[14] While the official groups' discourse provide evidence for the creation of the three categories, my conversations offer opportunities for exploring engagement strategies based on my typology and how official discourse is reflected or taken up by individuals.

Chapter outline

This book addresses the complicated relationship between Christianity and the environment and how it can result in climate skepticism but does not have to be a permanent roadblock to productive deliberation and advocacy. Before outlining strategies for addressing each type of response to environmentalism, I will first describe each of the categories in more detail using examples of their discourse. Each explanatory chapter makes use of exemplary groups and actors who primarily rely on strategies of separating, bargaining, and harmonizing. The strategy chapters build on the preceding explanatory chapters by directly linking rhetorical patterns with effective engagement, redirection, and refutation strategies. McGee (1990) argued that texts do not exist on their own, but it is up to the rhetorical critic to construct them from the variety of rhetorical texts and contexts that exist in the public. With this perspective in mind, I acknowledge that the fragments that I have assembled here create a case for the distinct, and rhetorical significant, discursive patterns that I argue are characteristic of separators, bargainers, and harmonizers. These fragments have been assembled from personal conversations coupled with the online publications of the exemplar groups, the CA, the AI, and the EEN, respectively. These fragments are by no means meant to be representative or generalizable to all religious climate change discourse. But, the fragments examined here do serve as examples of circulating discourse that intersects faith and environmental attitudes.

Chapter 1 details the rhetorical characteristics and argument resources of the separators. Separators engage in an aggressive rhetoric that evokes metaphors of war, positions themselves as gatekeepers for definitions and authority, and employs a hierarchy of humans over animals. Chapter 2 uses the theoretical foundations of Chapter 1 to propose strategies for engaging skeptics who make use of separators' rhetorical strategies. This chapter details three argument strategies to respond to these separator traits: asking questions instead of making assertions, offering premises and working with inductive logic, and finding points of commonality between the environment and their values. This chapter also synthesizes previously proposed strategies and why certain strategies may be unsuccessful for engaging separators.

Chapter 3 describes the defining characteristics of bargainers. As the second type of climate skeptics, bargainers engage in metaphors of revolution, leverage scientific standards to stall productive deliberation, and cherry-pick experts and evidence to support their arguments. This category of skeptic is not as aggressive as the separators and will often view themselves as more scientific than mainstream environmental science. The bargainers' main strategy is to borrow concepts and standards from environmental science but modify and cherry-pick them to promote skepticism toward the scientific consensus on climate change. Chapter 4 identifies the core strategies for engaging bargainers differently than separators. Because bargainers approach environmental concerns differently than separators, I propose that the strategies we use to engage them must also be modified. To address bargainers, we should use strategies that work within the bargainers' framework, acknowledge the value of revolution, and offer counterexamples to their assumptions about environmentalists and those they are launching a revolution against.

Chapter 5 examines the category of harmonizers who recognize climate change as a scientifically-legitimized reality but may not engage in climate advocacy. Although harmonizers are not climate skeptics, people who use harmonizing strategies can easily become members of a disinterested public that acknowledges climate change but does not act.[15] This chapter outlines how harmonizers rework traditional associations between Christianity and the environment but may still stall at the level of enacting change. Chapter 6 explores strategies for engaging harmonizers in concrete environmental activism. Because harmonizers accept mainstream environmental conclusions, the strategies for addressing harmonizers focus on turning belief into motivated action. These strategies engage the harmonizers' worldview by shifting from private to public activism, communicating urgency, and encouraging harmonizers to think globally. Instead of trying to shift, reframe, or reimagine values (as we do with separators and bargainers), the harmonizers are encouraged to envision themselves as powerful agents capable of enacting change, motivated by their faith and their belief in the current environmental crisis.

The concluding chapter summarizes the previous chapters and my arguments for both recognizing the multifaceted relationship between the Christian identity and environmental beliefs and the utility of using tailored strategies for different types of religious responses to the climate change controversy. This chapter also returns to my theoretical contributions to environmental communication and my pairing of qualitative interviews with rhetorical analysis. I summarize my argument about how discourse, when viewed as a dynamic, productive performance of identity, sheds new light on the nature of climate skepticism and how rhetorical theory and rhetorical listening can be integrated into environmental argument studies for a holistic view of public controversy that proposes practical solutions.

The final section of this book is an executive summary of the rhetorical strategies provided for engaging religious adherents in environmental topics. This chapter describes talking points, strategies for starting conversations and keeping them going, and brief justifications for the proposed strategies. This chapter is meant to be a quickly perusable reference guide about how to engage religiously-motivated skeptics (and people in general) in productive conversations, the importance of valuing people (even if we disagree), and how a willingness to engage can provide opportunities for education, enlightenment, and understanding.

Conclusion

In closing, this book analyzes the rhetoric of contemporary Christian discourse about the environment with the hope of responding to skepticism and apathy and of leveraging religious identity for productive environmental activism. Christianity, as a powerful meaning-maker and core identity in the United States, is often linked to climate change denial, skepticism, and inaction. But, I argue that religion can also be a part of an ecologically integrative and "restorative discourse" that advocates for climate change action (Milstein 2012: 162) with strategic pivoting and reframing. Through the combined power of rhetorical criticism and argument theory (Bloomfield and Tscholl 2018), I explore rhetorical pathways for reinterpreting the relationship between Christianity and the environment.

It is my goal to offer hope in the rhetorical pursuit of productive deliberation about the environment despite consistent public apathy and skepticism. Subsequent chapters will provide evidence that there are patterned argument strategies that religious adherents use in response to environmental information. By exploring the underlying frameworks that support their worldviews and mapping the arguments they make, we can put forth strategies for effective countering but also understanding, in the creation of a healthy, productive public sphere (Goodnight 1987). If we value public participation, silencing skeptics and not taking them seriously has the potential to irreparably harm the future of environmental communication. By changing the conversation, our language, and the frames we invoke, we can strive toward a public sphere unafraid to address environmental conversations and offer opportunities for cooperation toward shared environmental values.

Notes

1 Many scholars (e.g., Hart and Nisbet 2012; McFadden 2016) warn against relying on knowledge gaps as the primary explanation of climate change denial. To view climate change simply as a lack of information (i.e., the information deficit model) glosses over the influence that other factors, such as identity, motivated reasoning, personal experiences, etc. have on information processing. Moving past exclusive focuses on information is a promising route forward in climate change discourse.

2 There are complicated issues involved in using the term "skeptic." In my research and interviews, I find that people who doubt the scientific consensus on the effects/presence of climate change gravitate toward the skeptic label instead of the denier label. Both point toward the same opposition to mainstream environmental conclusions, but the skeptic label is often interpreted more neutrally or positively as it implies a natural, scientific focus on questioning assumptions, which scholars have previously argued is a common attribute of creationists, AIDS dissenters, as well as the current climate change population under discussion (Bloomfield and Tillery 2019; Ceccarelli 2011; Pennock 2003). For these reasons, I use the term "skeptic" as an umbrella term for both the separators and the bargainers.

3 Categories, by definition, create boxes through which we can learn about the world but that potentially limit the options available to see and understand. As I will discuss later, Burke (1969: 59), in addition to Ott and Dickinson (2013), warned that any "selections" of reality end up becoming "deflections" of other ways to view reality. It is my hope that this particular type of selection provides insight into the relationship between Christianity and the environment without discouraging different ways of selecting and reflecting on this relationship. I take these potential consequences very seriously and worked in this book to let categories of responses emerge organically by studying the discourse of a variety of groups.

4 This strategy has also been proposed by Cozen *et al.* (2018), Goodwin and Dahlstrom (2014), Hine *et al.* (2014), and Sangalang and Bloomfield (2018) for engaging in climate and environmental communication.

5 These typologies certainly tell us important information about the variety of ways that skeptics form arguments against mainstream environmental science, but I argue that there are alternative ways to conceptualize typologies to understand why people might react or respond to environmental information in different ways. In a sense, these typologies construct unique categories based on the consequences of skepticism (e.g., strength of denial, type of doubt, and features of environmental science that are causes for doubt), while my typology focuses on the frameworks and features that led to skeptical or apathetic attitudes toward the environment.

6 These scholars are too numerous to list here, even as an overview, but will be referenced as needed in the subsequent strategy chapters.

7 Many scholars have forwarded the idea of words and symbols having material effects. In addition to Burke, Susanne Langer (1974: 317) is an important theorist in conceptualizing how words are products of bodies and are "systems of feeling" between an individual and their environment (broadly conceived). Our words, therefore, are not meaningless or arbitrary choices, but are immensely consequential.

8 I sometimes refer to grounds as "inventional resources," because they serve as raw materials from which arguments are formed, the foundation from which people can make certain claims. The term "inventional" comes from the first of Cicero's canons of rhetoric, invention, which describes the discovery and finding of ideas.

9 "The Lord God took the man and put him in the Garden of Eden to work it and take care of it" (New International Version).

10 This consensus has been well established by a series of reports that analyze both abstracts of peer-reviewed scientific reports and interviews with experts in environmental science and climatology (e.g., Anderegg *et al.* 2010; Doran and Zimmerman 2009; Oreskes 2004). Of note is a study by Bray (2010) who offers that those who do dissent from Intergovernmental Panel on Climate Change (IPCC) reports may do so on the grounds that they disagree with the underestimation of IPCC data and not a disagreement with their general conclusions about the presence and severity of climate change.

11 Milstein *et al.* (2011: 488) argue that instead of saying humans and the environment and reifying the nature–culture binary, scholars should turn to "ecoculture" and "humanature" as terms that more accurately reflect through language the way "they are in life." I wholeheartedly agree with their assertions that attention to naming is of deep importance but choose to use the separated terms in this book to reflect how they appear in the discourse of my artifacts and dialogue partners.

12 These notes about labeling, and thus separating, will emerge as an important feature of the separators' discourse, who position human needs and environmental needs as separate, competitive, and a zero-sum game.

13 These verses will emerge frequently as inventional resources (or grounds) for individuals who invoke separator arguments. They are meaningful because they validate a hierarchical relationship where it is not only a good thing to focus on human needs, but a godly mandate.

14 My total data pool consists of 54 survey responses (from Bloomfield forthcoming), 12 phone interviews, and 36 online chat exchanges that spanned most of 2018. Phone interviews and online chat exchanges were arranged through solicitations for conversations posted to various Christian and climate change subreddits on Reddit.com. Respondents agreed to have their comments referenced in this research anonymously. When referencing conversations in the text, I use randomly assigned pseudonyms that are not representative of gender, race, or age.

15 The Yale Program on Climate Change Communication team has proposed a six-tiered system for analyzing how engaged people are in the topic of climate change (Roser-Renouf *et al.* 2016). As of their March 2016 study, only 17 percent of Americans fall into the "alarmed" category, meaning that they believe in global warming, are concerned, and are motivated to act on those beliefs. As one moves down the chart, belief, concern, and motivation wane, with the "concerned" category (28 percent of the population) believing in climate change, but not engaging it personally. My engagement with harmonizers leads me to believe that many would be members of the Yale Program's concerned group.

References

Adrián Villar Rojas: The Theater of Disappearance (n.d.) Available at: www.moca.org/exhibition/adrian-villar-rojas-the-theater-of-disappearance (accessed January 17, 2018).

Anderegg, W.R.L., Prall, J.W., Harold, J., and Schneider, S.H. (2010) Expert Credibility in Climate Change. *Proceedings of the National Academy of Sciences* 107(27): 12107–12109. DOI: 10.1073/pnas.1003187107.

Antilla, L. (2005) Climate of Scepticism: US Newspaper Coverage of the Science of Climate Change. *Global Environmental Change* 15(4): 338–352. DOI: 10.1016/j.gloenvcha.2005.08.003.

Barker, D.C. and Bearce, D.H. (2013) End-Times Theology, the Shadow of the Future, and Public Resistance to Addressing Global Climate Change. *Political Research Quarterly* 66(2): 267–279. DOI: 10.1177/1065912912442243.

Bauer, M.W., Allum, N., and Miller, S. (2007) What Can We Learn from 25 Years of PUS Survey Research? Liberating and Expanding the Agenda. *Public Understanding of Science* 16(1): 79–95. DOI: 10.1177/0963662506071287.

Bloomfield, E.F. (2017) Ark Encounter as Material Apocalyptic Rhetoric: Contemporary Creationist Strategies on Board Noah's Ark. *Southern Communication Journal* 82(5): 263–277.

Bloomfield, E.F. (forthcoming) Ecocultural Identity in the Creation Care Movement: Analyzing Contemporary Performance of Religious Environmentalism. In: Milstein, T. and Castro-Sotomayor, J. (eds.) *The Routledge Handbook of Ecocultural Identity*. New York; London: Routledge.

Bloomfield, E.F. and Lake, R.A. (2015) Negotiating the End of the World in Climate Change Rhetoric: Climate Skepticism, Science, and Arguments. In: Meisner, M.S., Sriskandarajah, N., and Depoe, S.P. (eds.) *Communication for the Commons: Revisiting Participation and the Environment*, pp. 384–396. Uppsala, Sweden: The International Environmental Communication Association.

Bloomfield, E.F. and Sangalang, A. (2014) Juxtaposition as Visual Argument: Health Rhetoric in Super Size Me and Fat Head. *Argumentation and Advocacy* 50(3): 141–156.

Bloomfield, E.F. and Tillery, D. (2019) The Circulation of Climate Change Denial Online: Rhetorical and Networking Strategies on Facebook. *Environmental Communication* 13(1): 23–34. DOI: 10.1080/17524032.2018.1527378.

Bloomfield, E.F. and Tscholl, G. (2018) Analyzing Warrants and Worldviews in the Rhetoric of Donald Trump and Hillary Clinton: Burke and Argumentation in the 2016 Presidential Election. *Kenneth Burke Journal* 13(2): n.p. Available at: http://kbjournal. org/analyzing_warrants_bloomfield_tscholl (accessed December 1, 2018).

Bray, D. (2010) The Scientific Consensus of Climate Change Revisited. *Environmental Science & Policy* 13(5): 340–350. DOI: 10.1016/j.envsci.2010.04.001.

Brockriede, W. (1974) Rhetorical Criticism as Argument. *Quarterly Journal of Speech* 60(2): 165–174. DOI: 10.1080/00335637409383222.

Brockriede, W. and Ehninger, D. (1960) Toulmin on Argument: An Interpretation and Application. *Quarterly Journal of Speech* 46(1): 44–53.

Brüggemann, M. and Engesser, S. (2017) Beyond False Balance: How Interpretive Journalism Shapes Media Coverage of Climate Change. *Global Environmental Change* 42: 58–67. DOI: 10.1016/j.gloenvcha.2016.11.004.

Burke, K. (1951) Rhetoric – Old and New. *The Journal of General Education* 5(3): 202–209.

Burke, K. (1966) *Language as Symbolic Action: Essays on Life, Literature, and Method*. Berkeley, CA: University of California Press.

Burke, K. (1968) *Counter-Statement*. First edition. Berkeley, CA: University of California Press.

Burke, K. (1969) *A Grammar of Motives*. Berkeley, CA: University of California Press.

Burke, K. (1974) *The Philosophy of Literary Form: Studies in Symbolic Action*. Berkeley, CA: University of California Press.

Burke, K. (1984) *Permanence and Change: An Anatomy of Purpose*. Berkeley, CA: University of California Press.

Call Out Climate Change Deniers (n.d.) Available at: www.ofa.us/climate-change-deniers/#/ (accessed January 17, 2018).

Ceccarelli, L. (2005) Let Us (Not) Theorize the Spaces of Contention. *Argumentation and Advocacy* 42(1): 30–33.

Ceccarelli, L. (2011) Manufactured Scientific Controversy: Science, Rhetoric, and Public Debate. *Rhetoric Public Affairs* 14(2): 195–228.

Ceccarelli, L. (2013) *On the Frontier of Science: An American Rhetoric of Exploration and Exploitation*. First edition. East Lansing: Michigan State University Press.

Clements, J.M., Xiao, C., and McCright, A.M. (2014) An Examination of the "Greening of Christianity" Thesis among Americans, 1993–2010: Green Christianity 1993–2010. *Journal for the Scientific Study of Religion* 53(2): 373–391. DOI: 10.1111/jssr.12116.

Coles, R.L. (2002) Manifest Destiny Adapted for 1990s' War Discourse: Mission and Destiny Intertwined. *Sociology of Religion* 63(4): 403–426. DOI: 10.2307/3712300.

Condit, C.M., Lynch, J., and Winderman, E. (2012) Recent Rhetorical Studies in Public Understanding of Science: Multiple Purposes and Strengths. *Public Understanding of Science* 21(4): 386–400. DOI: 10.1177/0963662512437330.

Conley, D.S. and Mullen, L.J. (2008) Righting the Commons in Red Rock Canyon. *Communication and Critical/Cultural Studies* 5(2): 180–199. DOI: 10.1080/14791420801989694.

Cozen, B., Endres, D., Peterson, T.R., Horton, C., and Barnett, J.T. (2018) Energy Communication: Theory and Praxis Towards a Sustainable Energy Future. *Environmental Communication* 12(3): 289–294. DOI: 10.1080/17524032.2017.1398176.

DeLuca, K.M. and Demo, A.T. (2000) Imaging Nature: Watkins, Yosemite, and the Birth of Environmentalism. *Critical Studies in Media Communication* 17(3): 241–260. DOI: 10.1080/15295030009388395.

Dickinson, G., Ott, B.L., and Aoki, E. (2006) Spaces of Remembering and Forgetting: The Reverent Eye/I at the Plains Indian Museum. *Communication and Critical/ Cultural Studies* 3(1): 27–47. DOI: 10.1080/14791420500505619.

Dickinson, G., Ott, B.L., and Eric, A. (2005) Memory and Myth at the Buffalo Bill Museum. *Western Journal of Communication* 69(2): 85–108. DOI: 10.1080/1057031 0500076684.

Domke, D. and Coe, K. (2010) *The God Strategy: How Religion Became a Political Weapon in America.* Oxford; New York: Oxford University Press. Available at: http:// books.google.com/books?hl=en&lr=&id=9co55qyB1lkC&oi=fnd&pg=PP9&dq=the+g od+strategy&ots=XhogVJGuIe&sig=J4jO20UOggOChZcwoS0FlTyP1mE (accessed November 11, 2012).

Doran, P.T. and Zimmerman, M.K. (2009) Examining the Scientific Consensus on Climate Change. *Eos, Transactions American Geophysical Union* 90(3): 22–23. DOI: 10.1029/2009EO030002.

Dorsey, L.G. (1995) The Frontier Myth in Presidential Rhetoric: Theodore Roosevelt's Campaign for Conservation. *Western Journal of Communication* 59(1): 1–19. DOI: 10.1080/10570319509374504.

Endres, D. (2009) Science and Public Participation: An Analysis of Public Scientific Argument in the Yucca Mountain Controversy. *Environmental Communication* 3(1): 49–75. DOI: 10.1080/17524030802704369.

Evans, J.H. (2018) *Morals Not Knowledge: Recasting the Contemporary U.S. Conflict between Religion and Science.* Oakland, CA: University of California Press.

Foss, S.K. (2017) *Rhetorical Criticism: Exploration and Practice.* Fifth edition. Long Grove, IL: Waveland Press.

Friedman, L. (2017) Syria Joins Paris Climate Accord, Leaving Only U.S. Opposed. *New York Times*, November 7. Available at: www.nytimes.com/2017/11/07/climate/syria-joins-paris-agreement.html (accessed January 31, 2018).

Gauchat, G. (2012) Politicization of Science in the Public Sphere: A Study of Public Trust in the United States, 1974 to 2010. *American Sociological Review* 77(2): 167–187. DOI: 10.1177/0003122412438225.

Goodnight, G.T. (1987) Public Discourse. *Critical Studies in Media Communication* 4(4): 428–432.

Goodnight, G.T. (2005) Science and Technology Controversy: A Rationale for Inquiry. *Argumentation and Advocacy* 42(1): 26.

Goodwin, J. and Dahlstrom, M.F. (2014) Communication Strategies for Earning Trust in Climate Change Debates. *Wiley Interdisciplinary Reviews: Climate Change* 5(1): 151–160. DOI: 10.1002/wcc.262.

Green, J.C. (2007) *The Faith Factor: How Religion Influences American Elections.* Westport, CT: Greenwood Publishing Group.

Hart, P.S. and Nisbet, E.C. (2012) Boomerang Effects in Science Communication: How Motivated Reasoning and Identity Cues Amplify Opinion Polarization about Climate Mitigation Policies. *Communication Research* 39(6): 701–723.

Hine, D.W., Reser, J.P., Morrison, M., Phillips, W.J., Nunn, P., and Cooksey, R. (2014) Audience Segmentation and Climate Change Communication: Conceptual and Methodological Considerations. *Wiley Interdisciplinary Reviews: Climate Change* 5(4): 441–459.

Hobson, K. and Niemeyer, S. (2013) "What Sceptics Believe": The Effects of Information and Deliberation on Climate Change Scepticism. *Public Understanding of Science* 22(4): 396–412. DOI: 10.1177/0963662511430459.

Hoerl, K.E. and Kelly, C.R. (2012) Genesis in Hyperreality: Legitimizing Disingenuous Controversy at the Creation Museum. *Argumentation and Advocacy* 48(3): 123–141.

Inhofe, S.J. (2012) *The Greatest Hoax: How the Global Warming Conspiracy Threatens Your Future.* First edition. Washington, DC: WND Books.

Intergovernmental Panel on Climate Change (IPCC) (2007) *IPCC Fourth Assessment Report: Climate Change 2007.* Geneva, Switzerland: Intergovernmental Panel on Climate Change. Available at: www.ipcc.ch/publications_and_data/publications_ipcc_fourth_assessment_report_synthesis_report.htm (accessed December 1, 2018).

Jacques, P.J., Dunlap, R.E., and Freeman, M. (2008) The Organisation of Denial: Conservative Think Tanks and Environmental Scepticism. *Environmental Politics* 17(3): 349–385. DOI: 10.1080/09644010802055576.

Jenkins, E.S. and Cisneros, J.D. (2013) Rhetoric and This Crazy Little "Thing" Called Love. *Review of Communication* 13(2): 85–107.

Jensen, R.E. (2016) *Infertility: Tracing the History of a Transformative Term.* RSA Series in Transdisciplinary Rhetoric. University Park, PA: Penn State Press.

King James Version. (n.d.) Available at: www.biblegateway.com/ (accessed December 1, 2018)

Kirk, K. (2018a) Climate Change Science Comeback Strategies. Yale Climate Connections, July 26. Available at: www.yaleclimateconnections.org/2018/07/climate-change-science-comeback-strategies-part-one/ (accessed December 19, 2018).

Kirk, K. (2018b) Focus on Those with an Open Mind. Yale Climate Connections, November 19. Available at: www.yaleclimateconnections.org/2018/11/focus-on-those-with-an-open-mind/ (accessed December 14, 2018).

Konisky, D.M. (2018) The Greening of Christianity? A Study of Environmental Attitudes over Time. *Environmental Politics* 27(2): 267–291.

Kortenkamp, K.V. and Basten, B. (2015) Environmental Science in the Media: Effects of Opposing Viewpoints on Risk and Uncertainty Perceptions. *Science Communication* 37(3): 287–313. DOI: 10.1177/1075547015574016.

Lakoff, G. (2010) Why it Matters How We Frame the Environment. *Environmental Communication* 4(1): 70–81. DOI: 10.1080/17524030903529749.

Lakoff, G. and Johnson, M. (1980) Conceptual Metaphor in Everyday Language. *The Journal of Philosophy* 77(8): 453–486.

Langer, S.K. (1974) *Mind: An Essay on Human Feeling.* Baltimore, MD: John Hopkins University Press.

Leiserowitz, A.A., Maibach, E., Roser-Renouf, C., and Smith, N. (2011) *Global Warming's Six Americas.* New Haven, CT: Yale Project of Climate Change, Yale School of Forestry & Environmental Studies and the Center for Climate Change Communication, George Mason University. Available at: http://environment.yale.edu/climate/files/SixAmericasMay2011.pdf (accessed April 6, 2013).

Lipka, M. (2016) U.S. Religious Groups and Their Political Leanings. Pew Research Center, February 23. Available at: www.pewresearch.org/fact-tank/2016/02/23/u-s-religious-groups-and-their-political-leanings/# (accessed January 17, 2018).

Long, R. (1983) The Role of Audience in Chaim Perelman's New Rhetoric. *Journal of Advanced Composition* 4: 107–117.

Matthews, P. (2015) Why Are People Skeptical about Climate Change? Some Insights from Blog Comments. *Environmental Communication* 9(2): 153–168. DOI: 10.1080/17524032.2014.999694.

McBurney, J.H. (1936) The Place of the Enthymeme in Rhetorical Theory. *Communications Monographs* 3(1): 49–74.

McCammack, B. (2007) Hot Damned America: Evangelicalism and the Climate Change Policy Debate. *American Quarterly* 59(3): 645–668.

McFadden, B.R. (2016) Examining the Gap between Science and Public Opinion about Genetically Modified Food and Global Warming. *PLOS ONE* 11(11): e0166140. DOI: 10.1371/journal.pone.0166140.

McGee, M.C. (1990) Text, Context, and the Fragmentation of Contemporary Culture. *Western Journal of Speech Communication* 54(3): 274–289. DOI: 10.1080/10570319 009374343.

McGeough, R.E., Palczewski, C.H., and Lake, R.A. (2015) Oppositional Memory Practices: U.S. Memorial Spaces as Arguments over Public Memory. *Argumentation & Advocacy* 51(4): 231–254.

Meyer, M. (2012) Aristotle's Rhetoric. *Topoi* 31(2): 249–252.

Miller, L. (2017) As "Climate Change" Fades from Government Sites, a Struggle to Archive Data. *Frontline*, December 8. Available at: www.pbs.org/wgbh/frontline/article/as-climate-change-fades-from-government-sites-a-struggle-to-archive-data/ (accessed January 31, 2018).

Milstein, T. (2011) Nature Identification: The Power of Pointing and Naming. *Environmental Communication* 5(1): 3–24.

Milstein, T. (2012) Greening Communication. In: Fassbinder, S.D., Nocella, A.J., and Kahn, R. (eds.) *Greening the Academy: Ecopedagogy through the Liberal Arts*, pp. 161–173. New York: Springer.

Milstein, T. and Dickinson, E. (2012) Gynocentric Greenwashing: The Discursive Gendering of Nature. *Communication, Culture & Critique* 5(4): 510–532.

Milstein, T., Anguiano, C., Sandoval, J., Chen, J.-W., and Dickinson, E. (2011) Communicating a "New" Environmental Vernacular: A Sense of Relations-in-Place. *Communication Monographs* 78(4): 486–510.

Mol, H. (1977) *Identity and the Sacred: A Sketch for a New Social-Scientific Theory of Religion*. New York: Free Press.

Mufson, S. (2017) Rick Perry Just Denied That Humans Are the Main Cause of Climate Change. *Washington Post*, June 19. Available at: www.washingtonpost.com/news/energy-environment/wp/2017/06/19/trumps-energy-secretary-just-denied-that-man-made-carbon-dioxide-is-the-main-driver-for-climate-change/ (accessed January 31, 2018).

Musser, M. (2013) Getting Gored: Michael Mann's Apocalyptic Prophecies and Al Jazeera's Green Jihad. *American Thinker*, February 19. Available at: www.Americanthinker.com/articles/../2013/02/getting_gored_michael_manns_apocalyptic_prophecies_and_al_jazeeras_green_jihad.html (accessed February 22, 2013).

New International Version. (n.d.) Available at: www.biblegateway.com/ (accessed December 1, 2018)

O'Leary, S.D. (1993) A Dramatistic Theory of Apocalyptic Rhetoric. *Quarterly Journal of Speech* 79(4): 385–426. DOI: 10.1080/00335639309384044.

Oreskes, N. (2004) The Scientific Consensus on Climate Change. *Science* 306(5702): 1686–1686. DOI: 10.1126/science.1103618.

Ott, B.L. and Dickinson, G. (2013) *The Routledge Reader in Rhetorical Criticism*. New York; London: Routledge.

Pennock, R.T. (2003) Creationism and Intelligent Design. *Annual Review of Genomics and Human Genetics* 4(1): 143–163.

Pew Research Center (2009) Section 5: Evolution, Climate Change and Other Issues. Center for the People and the Press, July 9. Available at: www.people-press.

org/2009/07/09/section-5-evolution-climate-change-and-other-issues/ (accessed August 9, 2013).

Pew Research Center (2013) GOP Deeply Divided Over Climate Change. Center for the People and the Press, November 1. Available at: www.people-press.org/2013/11/01/gop-deeply-divided-over-climate-change/ (accessed January 31, 2018).

Poortinga, W., Spence, A., Whitmarsh, L., Capstick, S., and Pidgeon, N.F. (2011) Uncertain Climate: An Investigation into Public Scepticism about Anthropogenic Climate Change. *Global Environmental Change* 21(3): 1015–1024. DOI: 10.1016/j.gloenvcha.2011.03.001.

Prelli, L.J. (1989) *A Rhetoric of Science: Inventing Scientific Discourse*. Columbia, SC: University of South Carolina Press.

Prelli, L.J. and Winters, T.S. (2009) Rhetorical Features of Green Evangelicalism. *Environmental Communication: A Journal of Nature and Culture* 3(2): 224–243. DOI: 10.1080/17524030902928785.

Ratcliffe, K. (1999) Rhetorical Listening: A Trope for Interpretive Invention and a "Code of Cross-Cultural Conduct." *College Composition and Communication* 51(2): 195–224. DOI: 10.2307/359039.

Ratcliffe, K. (2005) *Rhetorical Listening: Identification, Gender, Whiteness*. Books by Marquette University Faculty. Carbondale, IL: Southern Illinois University Press. Available at: https://epublications.marquette.edu/marq_fac-book/273 (accessed December 1, 2018).

Rickards, L.A. (2015) Metaphor and the Anthropocene: Presenting Humans as a Geological Force. *Geographical Research* 53(3): 280–287. DOI: 10.1111/1745-5871.12128.

Roser-Renouf, C., Maibach, E., Leiserowitz, A.A., and Rosenthal, S. (2016) Global Warming's Six Americas and the Election, 2016. Yale Program on Climate Change Communication, July 12. Available at: http://climatecommunication.yale.edu/publications/six-americas-2016-election/ (accessed May 3, 2018).

Ross, D.G. (2013) Common Topics and Commonplaces of Environmental Rhetoric. *Written Communication* 30(1): 91–131. DOI: 10.1177/0741088312465376.

Rushing, J.H. (1983) The Rhetoric of the American Western Myth. *Communication Monographs* 50(1): 14–32. DOI: 10.1080/03637758309390151.

Sangalang, A. and Bloomfield, E.F. (2018) Mother Goose and Mother Nature: Designing Stories to Communicate Information about Climate Change. *Communication Studies* 69(5): 583–604.

Smith, G.A. and Martinez, J. (2016) How the Faithful Voted: A Preliminary 2016 Analysis. Pew Research Center, November 9. Available at: www.pewresearch.org/fact-tank/2016/11/09/how-the-faithful-voted-a-preliminary-2016-analysis/ (accessed January 17, 2018).

The Climate Change Deniers in Congress (n.d.) Available at: https://motherboard.vice.com/en_us/article/pg5zqg/a-guide-to-the-climate-change-deniers-in-congress (accessed January 17, 2018).

Tillery, D. (2018) *Commonplaces of Scientific Evidence in Environmental Discourses*. First edition. New York; London: Routledge.

Waddell, C. (1990) The Role of Pathos in the Decision-Making Process: A Study in the Rhetoric of Science Policy. *Quarterly Journal of Speech* 76(4): 381–400. DOI: 10.1080/00335639009383932.

White Jr., L. (1967) The Historical Roots of Our Ecological Crisis. *Science* 155(3767): 1203–1207.

1 Defining the separators

War, melodrama, and authority

Rabbi Dr. Abraham Twerski tells a parable called "fish love" about two people discussing the love they have for the things that they consume. Upon seeing the first eat a fish and remark, "I love fish," the other responds that their love for fish is not a true, giving love, but an internal, selfish love based on how the fish fulfills that person's needs (BHAN 2017). While true love is reciprocal and offers genuine care for the person/thing being loved, "fish love" is a phrase synonymous with exploitative, manipulative love where only one side of the relationship benefits from the interaction. The one-sided relationship of "fish love" is one way to understand conceptions of nature as a tool to be exploited for the benefit of humanity (Milstein and Dickinson 2012). While people might note that they care for and love the environment, they may, in practice, not value nature beyond its capabilities for human survival and enjoyment.

Fish love is a useful lens through which to view separators, who may profess care for the environment, but often define the environment's value only in terms of its utility for humans. Separators thus evoke a hierarchy of humans over nature and use that hierarchy to guide their orientation toward proper environmental actions. This hierarchy is in part validated by certain Christian interpretations that laud humans as the pinnacle of God's creation. The separators are so named because their primary strategy for engaging environmental topics is to cleave their religious identity from an environmental one. Environmental identities, or "ecocultural" identities, are rarely cultivated in a vacuum. Instead, they emerge through the intersection and overlap of other identities such as political, religious, economic, and social ones.

An important point to consider throughout this book, and one that I am re-emphasizing here, is that religious adherents are not "anti-science" and do not participate in "science denial" for all meanings of the word science. For example, religious adherents make use of modern technology, adopt basic scientific principles, and often turn to science to explain many phenomena in their lives (Evans 2018). When I discuss how separators create distinctions between their religious interpretations and "science," I am using the term to refer to mainstream climate science that supports environmental advocacy. To say that separators, or religious climate skeptics, enact the war between religion and all of science is an overstatement of their particular arguments, ideologies, and identities. It is more accurate to say that the war the separators are launching is a figurative counter-attack to what they view as moral failings of climate science and as negative consequences of enacting environmentally-oriented policies.

In the discourse examined below, I trace separation as a rhetorical pattern that underlies a polarized, aggressive form of climate change denial. Separators create distinct rhetorical boundaries between friend and foe, making it difficult to open conversation when one is perceived as being on the "side" of the environment. This worldview potentially situates others as enemies who are not to be trusted, making productive dialogue challenging. Indeed, to engage with separators is an attempt to address a deep-seated rejection of mainstream interpretations of climate change. In order to create opportunities for conversation and to explore potential avenues for rhetorical correction, it is first integral to understand the frames, metaphors, and narratives that undergird separators' understanding of the relationship between their faith and the environment.

In what follows, I analyze the discursive "fragments" of separators, from the Cornwall Alliance (CA) and personal conversations, to show their shared characteristics, rhetorical patterns, and argument strategies (McGee 1990: 279). I argue that the separators primarily make use of three argument strategies: (1) controlling definitions, (2) shifting the blame onto the environmental movement, and (3) appealing to the Bible's authority to construct value hierarchies. The separators' discourse features metaphors of war that can foster hostility and cynicism toward environmentalism and the topic of climate change. First, I describe the frames and metaphors of war and hostility in more detail, establishing both the foundation of the separators' argument strategies and prefacing what distinguishes them from bargainers. Then, I use examples from publically available discourse and publications from the CA and interpersonal conversations to analyze the three argument strategies that emerge from the separators' war frame. I identify these argument forms to illuminate the distinct rhetorical features of the separators and to provide resources from which to engage separators and their arguments in the following chapter.

The Cornwall Alliance

Started in 2005 as the Interfaith Stewardship Alliance, the CA is an evangelical group that supports conservative environmental policies. The CA's slogan is "for the stewardship of creation," indicating a concern for the protection of God's creation, but we will see that this term emphasizes the comfort and protection of human life (CA 2018: para. 1). The CA borrows the term "stewardship" from the Creation Care movement, which uses religion to advocate for environmental protection. But, the CA's definition of stewardship leads them to conclude that pro-environmental policies are dangerous to religion and society. Specifically, the CA (2013: para. 1) advocates for "economic development built on Biblical principles" and argues that excessive environmental protections inhibit economic progress, harming humans in the process. Dr. E. Calvin Beisner is the founder of the CA and publishes most of the content on the CA's website. In addition to publishing articles, the CA hosts events, provides consulting services, and educates the public about the dangers of the environmental movement. Included among their publications are moral declarations, open letters, and responses to what the CA considers scientifically-biased news and scholarly articles. The CA seeks to validate its Christian perspective and advocate for using the Bible to make decisions on environmental policy.

Some scholars have cautioned against focusing on individual groups, such as the CA, as exemplars of religious discourse (Evans 2018). To avoid such reductionism, I pair an analysis of the CA's discourse with individuals who share these perspectives to get a fuller picture of the influence and circulation of war metaphors in religious climate change rhetoric. Furthermore, I justify focusing on the CA's rhetoric because the CA is an important player in the climate change controversy, where traces of its discourse reverberate through the public sphere. For example, the CA regularly publishes public commentaries, gives lectures at conferences hosted by anti-environmental groups such as the Non-governmental International Panel on Climate Change, and has released a popular video series, *Resisting the Green Dragon* (Beisner 2011). The CA frequently collaborates with the Heartland Institute and the Heritage Foundation, prominent conservative think tanks that advance economic arguments against environmental protection. The CA was in the news for supporting Trump's appointment of Scott Pruitt to the Environmental Protection Agency (EPA) by sending a letter with the signatures of scientists who supported the decision to Congress (Beisner 2017c; Leavenworth 2017).

Due to the CA's extensive funding by and connections to the fossil fuel industry, some have questioned the sincerity of the CA's religious arguments against climate change (O'Connor 2017). It is quite possible that in the CA conservative industry leaders have found compatriots, fighting against environmental policies for vastly different reasons. Whether motivated solely by faith or also by financial interests, the CA fosters and circulates aggressive, belligerent attitudes toward the environment, which get picked up and repeated in online spaces. O'Connor (2017: para. 13) argued that the CA is akin to a religious "enforcer" that keeps the Christian, and specifically evangelical community in line in order to stop environmental attitudes from "gain[ing] a following" within Christianity. Studies have shown that Christians are not, on the whole, "going green" or expressing more environmentally-friendly attitudes (Konisky 2018), so perhaps the CA and other groups have played a role in effectively quashing environmental activism in the community.

Given the reach and public presence of the CA and their primary use of aggressive discourse often referenced in climate communication studies (Evans 2018), I have selected them to be the exemplar separator. It is important to remember that these categories emerged organically from reading publications, news articles, and in interviews with climate skeptics, and I now describe prominent rhetorical characteristics with examples that I located within that examined discourse. Before elaborating on the primary features of the separators' discourse, I analyze their discursive framework that hangs on metaphors of war, hostility, and separation.

Guiding terms

The separators' guiding terms serve as the terministic screen, or filter, through which they view the climate change controversy (Burke 1966: 45). It is important not to view these guiding terms as deterministic or exhaustive, but as important features of the separators' environmental worldview that influence how they perceive mainstream climate science, engage in conversation, and prioritize environmental topics. In viewing mainstream environmentalism as a challenge to their faith, separators build symbolic walls to protect their values and morals. The CA's discourse is quite extreme in part because they perceive that climate science, which supports environmental policies, has been tainted by undue liberal and secular influences and thus should be avoided and destroyed. I do not claim that the CA's discourse is representative of all climate deniers. Indeed, the CA do not consider themselves deniers, but that they have the correct scientific and Biblical interpretation of appropriate human behavior toward the environment. The CA is a particular group of climate skeptics that injects public conversation with aggression and hostility, protecting their faith and values against what they perceive to be malicious environmental influences.

I argue that the CA's discourse is an important contributor to the climate conversation because it legitimizes aggressive inventional resources that are echoed by politicians, leaders, and publics. These attitudes and beliefs get taken up in online spaces, as well, where people who distrust climate science and turn to their faith for guidance on environmental matters and may find resonance with the CA's messages. I support this argument by detailing conversations I have had with religious climate deniers on online discussion forums about their beliefs. Previous studies on climate denial discourse online has found that skeptics sometimes turn to insults (Ross 2013) and threats of violence (Bloomfield and Tillery 2019) as appropriate ways to engage environmental topics, providing evidence that aggression is an active part of climate change discourse.

Separators highlight a distinct subset in the constellation of climate discourse that uses rhetoric to encourage aggressive attitudes toward environmentalism and foster fears of the environmental movement's consequences. A faith-based frame of war constructs opposing opinions on climate science as a holy war. In a holy war, there is no room for gray areas or compromise. Everyone who is not fully aligned with the separators' beliefs is deemed untrustworthy. Given this attitude, it is easy to see how working within a framework of war might inhibit productive conversation, which relies on the willingness of multiple parties to engage and respect one another. In addition, the war framing raises the stakes of the conversation, where separators believe there is much more on the line that simply a belief in climate change, but also one's identity as a Christian (McCammack 2007). From this perspective, separators' rhetorical strategies are far more than counter-arguments to what they view as harmful environmental policies; they are also performances of their faith. When climate science appears to contradict Christian practices by seemingly promoting the pagan worship of the Earth or valuing the environment over vulnerable human life, separators respond with a vocal defense of their moral priorities.

The metaphor of war influences separators' attitudes and behaviors in prominent ways. First, separators create distinct lines between those they perceive to be correctly following the Bible and those who turn against it. In viewing climate science through the lens of Christianity, the Bible can be considered a "written Constitution" or originating terminology that prescribes certain actions and outlines consequences for deviating from its rules (Burke 1969: 377). To reject mainstream climate science in favor of literal readings of the Bible signals a move in favor of one side over the other. Because climate science is perceived to begin from a different starting point outside of the Bible, separators sometimes label mainstream climate science as anathema to the flourishing of the Christian worldview. After enemies have been identified, there is a clear and simple route forward: eliminate them and thus end the war. Separators thus enact a clear separation between true heroes and evil villains, which is easy to understand and envision. The "clarity and simplicity in assigning wrong predominantly to one side" offered by war frames makes the frame particularly appealing, attention-grabbing, and meaningful (Desilet and Appel 2011: 348). In this sense, the use of the war metaphor may be a purposeful strategy to capture media attention hungry for polarized narratives (Dixon and Clarke 2013) and to mobilize those who share the separators' morals and values to take the potential threats of environmentalism more seriously.

It is important to note that launching counter-arguments against environmentalism is not viewed as a neutral act, but establishes that one side represents clear "*moral* wrongs [and] injustices that cannot be rectified through political compromises or minor adjustments" (Schwarze 2006: 250, emphasis in original). Separators may view discourse and discussion of the environment as more than a simple verbal exchange, but as direct attacks on their worldview and faith. The separators' rhetoric and behaviors, therefore, reflect "an arduous heroic journey" that must be undertaken until "the moral balance is restored" (Lakoff 1991: 30). In restoring that balance, Lakoff (1991: 30) argued that the heroes of the story "cannot negotiate with villains; they must defeat them." In response to their enemies' attacks, separators position themselves as the defenders of their faith and the heroes of the war, solidifying the separation between themselves and their perceived enemies. The adoption of a war frame also serves a legitimizing function in the separators' arguments by which an ongoing war validates aggressive argument strategies in response to mainstream environmentalism. Steinert (2003: 266) argued that "war forges the bond of community ... like nothing else," because war clearly sets a distinction between opposing sides. In adopting a war metaphor, separators legitimize aggressive responses toward environmentalists in defense of their faith.

This war frame is most reminiscent of the genre of melodrama, through which strict, dividing lines are drawn, differences amplified, and stakes raised. Schwarze (2006: 244) argued that melodramas can be productive in environmental conversations in that they work to "stag[e] new visions of moral order" against one deemed immoral, restrictive, and oppressive. Melodramas thus encourage an "emotional identification with victors or victims, whether celebrating the former or sympathizing with the latter" and provide a "motive force for collective action" to restore the lost order (Schwarze 2006: 244). By invoking a metaphor of war, the separators create a "connect[ion] to strong emotions and social values" that work to "mobiliz[e] people and resources" toward their perspective (Schwarze 2006: 268). While Schwarze (2006) engaged the opportunities for melodrama in environmental discourse, melodrama emerges in separators' discourse as a way to shut down productive environmental communication and to polarize the controversy between warring sides. The use of war metaphors as part of the separators' vocabulary stems from their melodramatic terministic screen and highlights the need for people to identify which side they are on: the side of the heroes (separators) or the many villains that challenge them.

While I use the term *strategies* to describe the following three discursive features of separators' rhetoric, I do not wish to propose that these features are necessarily or perpetually intentional. Instead, these features are logical extensions of the separators' guiding framework of war and melodrama that center on hostility toward environmentalism and a fierce desire to protect their morality and values against perceived threats. Because of this framework, I argue that the separators' discourse features rhetorical patterns of controlling definitions, shifting blame, and appealing to authority. These three features illuminate the ways that the guiding terms of war are performed in separators' discourse. By attending to these features, we can locate opportunities for engagement and refutation in the next chapter.

Controlling definitions

Burke (1966) attributes much power and influence to our vocabulary choices, which make up the terministic screens through which we view the world. To make a choice between two words is not an arbitrary decision, but one indicative of one's perspective on a situation. In making a choice to engage the issue of climate change as one of aggression, defense, and hostility, separators put themselves in the position to defend their faith, values, and morals against insidious forces. Part of this defense strategy is to control labels and the membership status that comes with those labels. Lake (1997: 67, 66) argued that identities are constructed "continually within and against," meaning that identity comes from the performance of group values and beliefs and also from separating those values from an "Other." In conceptualizing identity as an argument between oppositional tensions (Bloomfield 2018; Lake 1997), we open ourselves up to considering the rhetorical dynamics present in religious environmental discourse. This section focuses on three definitions that separators control by drawing strict boundaries and criteria: engaging in "stewardship," being a good "Christian," and being a real "scientist."

Religious environmental groups that advocate for environmental protection often link their environmental attitudes to a moral obligation to care for the

Earth, called stewardship (Bloomfield forthcoming). While I will expand more on how harmonizers make use of the term stewardship in Chapter 5, it is sufficient for now to know that the term is used in support of pro-environmental discourse. The CA, arguably emerging in response to the Creation Care movement, co-opts the term in their slogan: "For the Stewardship of Creation" (CA 2018). By using "stewardship," the CA presents itself as part of religious environmentalism. The CA would argue that they do support the environment but they point to the Bible for justification as to the best way to protect it, which differs from mainstream climate science's recommendations. The CA certainly plays up a pro-environment appearance on its website, which loads with a bright green and blue background with a negative-space tree built into their logo. But, in using the term stewardship, the CA does not wish to evoke a Creation Care identity. Instead, it promotes a new definition of stewardship that privileges people and markets over nature. This definition of stewardship evokes a hierarchy that positions the environment in a "lower," subordinate position to human needs (Milstein and Dickinson 2012: 526). The CA's interpretation of stewardship serves as a gatekeeping device to separate "true" Biblical stewardship from what an online conversant called perverse, "Gaia-worshipping" interpretations of environmentalism.

On their "What We Do" page, the CA (n.d.-f: para. 2) defines "Biblical earth stewardship" as "godly dominion," where "men and women [work] together to enhance the fruitfulness, beauty, and safety of the earth to the glory of God." In this definition, we see a qualification of the idea of "stewardship" with "Biblical earth," reiterating the role of the Bible in providing the correct definitions, and the addition of "dominion" as a synonym for stewardship. It is also interesting to note that this definition places humans in an active role of intervening in the environment by "working" it only for the benefit of humans and God, instead of a stewardship focused on preservation and non-interference in natural flourishing. The CA warns that Christians "must never conflate Biblical earth stewardship with environmentalism. The two are mutually exposed from – pardon the pun – the ground up" (Beisner 2015a: para. 1). The CA severs environmentalism from Christian values, placing them in antagonistic positions that do not and should not blend. The CA argues that environmentalists take their cues from pro-environment climate science instead of the Bible and thus, by definition, enact a non-Biblical definition of stewardship.

In a conversation with a separator online I call Liam, he described his perspective on climate change as "responsible stewardship." When asked to elaborate what he meant by that term, Liam simply noted that practicing responsible stewardship would likely not include "sacrifice or inconvenience" for humans, as that would be "irresponsible" to other humans. Similar to the CA's discourse, Liam was wary of making changes to his everyday life as he valued his own wellbeing (and the wellbeing of other humans) over that of the environment. Liam continued, "This is our (Humanity's) planet, we have dominion" and have the right to "liv[e] a fulfilling life." These beliefs can be traced to the separators' faith, as they stem from the dominion mandate of Genesis 1:26.[1] Seen in both

online conversations and through the discourse of the CA, the separators' definition of stewardship and proper care for the environment prioritizes human
comfort and survival. The different interpretations of stewardship help separators distinguish between those who are following the Bible as inspiration for
their environmental attitudes and those who turn elsewhere. For the CA, to
engage in the environmental movement's form of stewardship is to compromise
one's status as a good Christian who is following proper Biblical stewardship.

The primary identity for separators is their faith, so determining who is Christian and who is not sets up an easy divide between friend and foe. Controlling
the definition of "Christian" created a self-fulfilling prophecy by which people
who believe and act differently from the separators disqualify themselves as true
Christians. Choosing to engage in certain activities becomes a constitutive
process through which an identity is formed. In other words, one is not born a
Christian, but becomes one through their behaviors and beliefs. A Christian may
label themselves as such, but without adhering to the proper actions and attitudes
of a Christian (as defined by the separator), one does not deserve the label.

For example, in another online conversation with Olivia, she defined being a
"true" Christian as starting with the Bible on all topics. She wrote, "If it looks
like a Christian, talks like a Christian, but doesn't point to the cross then it's not
a Christian, it's something else." For Olivia, if faith was not the utmost priority
or standard by which a Christian acted at all times, they were not a true Christian, despite what they may call themselves. Prioritizing the environment over
faith (or appearing to) can be interpreted by separators to be a betrayal of the
faith. Upon questioning the separator further on the importance of labels and
how she thought a person who claimed to be a Christian might not be, Olivia
responded that "terms are important because truth is important." Without the
right terms, the right labels, this person felt that calling oneself a Christian was
meaningless; it was only through a particular interpretation of faith (one that
filters everything through the Bible) that one could distinguish real and true
Christians from charlatans. "Christians" who used the label but differed from
Olivia's interpretation were not viewed in a positive light and made her even
more skeptical of Christian environmentalists.

The CA shares this dialogue partner's view of compromising Christians,
arguing that pro-environmental Christians reinterpret and distort the Bible to
their environmental will. Christian environmentalists "undermine the message of
Genesis," because they "misinterpret" key Bible verses by "borrowing" from the
environmental "worldview" (Beisner 2012: para. 20). In its statement of faith,
the CA describes the environmental movement as "rest[ing] on poor theology,
with a worldview of the Earth and its climate system contrary to that taught in
the Bible" (2009b: 1). Despite groups calling themselves Christian, the CA
doubts their true Christian status because they are perceived to have separated
themselves from the Bible's teaching. Simply put, the CA describes the environmental movement as "overwhelmingly anti-Christian," thus cleaving a Christian
identity from environmentalism (CA n.d.-f: para. 3). Similar to definitions of
stewardship, the CA constructs a chasm between a "true" Christian identity, as

defined by separators, and environmentalism, thus arguing that one can be one or the other, but never both.

The inverse of this Toulmin model implies that those who are moral and perform moral acts could not themselves be environmentalists. In other words, people who profess to believe in Christ thus are not truly Christian if they are also environmental advocates; they are mutually exclusive identities. The separators' strict conception of identity is rooted in the melodramatic narrative of war they construct, where if one is not good, one must be evil. This conditional syllogism can be understood as such:

The Cornwall Alliance argument about "true" Christians

Premise 1: True Christians are good, moral, and believe in Christ.
Premise 2: Environmentalists are evil, immoral, and abandon Christ's teachings.
Conclusion: Environmentalists cannot be "true" Christians.

The CA frequently targets specific Christian organizations, such as the Evangelical Environmental Network (EEN, which I analyze in Chapter 5 as the exemplar harmonizer group) as compromising God's word. The CA describes groups such as the EEN as being motivated to protect the environment because of their "mistaken understandings" of what the Bible says about the relationship between humanity and the Earth (Beisner 2012: para. 23). To interpret the Bible differently from the CA is no minor issue; it separates true Christians from fraudulent Christians. In the CA's narrative, Christian environmentalists are guilty of both "green hypocrisy" and "poor theology," because they neither conform to the CA's interpretation of proper stewardship nor to the Bible (Beisner 2017c; CA 2009a: 1).

The CA also doubts that harmonizing Christians are true Christians because of their perceived political leanings. The CA describes itself as "up against a well-funded movement, one that receives millions of dollars from foundations whose agendas are diametrically opposed to the fundamental Christian ethic of the sanctity of human life" and the protection of the poor (Beisner 2015b: para. 11). The CA portrays itself as an underdog that is facing a daunting battle, evoking a war frame where the group and those who share the CA's beliefs are under attack from a more formidable foe that has immoral financial baggage. In a report about donations the EEN receives, Beisner (2015b: para. 2) argued that the group collects funds from "Left-wing, pro-abortion, pro-population control, environmentalist foundations." These ideologies are viewed as antithetical to the more conservative social positions the CA takes toward abortion and the economy. Beisner (2014: para. 1) accuses the EEN and similar groups of enacting what he calls "environmental deceit" by favoring the health of the environment at the risk of harming the pro-life movement and sacrificing traditional, Christian social values.

Along with damage to the pro-life movement and conservative social values, the CA is particularly concerned about the effects that environmentalism will have on poor communities:

Preventing climate change does not help the poor, it dooms them! Poverty simply kills more people than climate. Consequently, it would be immoral to deny the poor the ability to develop by curtailing their access to abundant, affordable, reliable energy, all in pursuit of an environmental objective that only interests one billion rich people.

(CA 2015: para. 20)

The CA equates environmentalism with harm to the poor and harm to the poor as immoral. Mapped as a series of premises, the deductive syllogism looks like this:

The Cornwall Alliance argument about the poor

Major premise: We should not do harm to the poor.
Minor premise: Environmentalism harms the poor.
Conclusion: We should not support environmentalism.

This argument model can be easily adapted by filling in "harm to the poor" with any Christian value, leading to the same conclusion that "We should not support environmentalism" as long as the minor premise is established through evidence or assertion. Indeed, we might consider the major premise of many of the CA's anti-environmentalist arguments to stem from iterations of "We should not commit sin or be immoral" and the minor premise becomes the association of the environmental movement with those named sins.

The rhetorical reasoning behind identifying fellow Christians as enemies can be explained by Burke's ideas of piety and order. Burke called piety a "system-builder" through which life experiences become "a unified whole" (Burke 1984: 75). Even a slight deviation from the order creates impiety and guilt, which undermines and destabilizes the entire order. As the saying goes, one cannot be a little pregnant; piety is an uncompromising identity. Likewise, separators view the Christian identity as a pure, holistic one where the slightest disobediences are tantamount to egregious ones. Bellah (1991: 218) argued that punishment levied on community violators echoes this distinction: "Failure to subscribe to [a faith's] beliefs is to be punished by banishment; falsely subscribing to them is punishable by death." To compromise, distort, and betray the faith is often considered a worse transgression than simple ignorance or rejection. Distortion implies an intentional, specific action that betrays the established order. In the separators' melodramatic war, there is no room for blurring or compromising; they have firm, literal interpretations of appropriate actions and behaviors and are exceedingly skeptical and even hostile to those that transgress. Such hostility is certainly understandable if one remembers the separators' melodramatic framework that ascribes evil to perceived opponents.

Christians and religious adherents are frequently mislabeled as anti-science (Evans 2018), when it is more accurate to describe them as skeptical or antagonistic toward a specific application or implication of science. For example,

Beisner (2010: para. 2) argued, "it becomes increasingly clear that a great deal of what's been called 'climate science' isn't science at all. It's ideological propaganda, often religious (but certainly not Biblical) masquerading as science." In addition to undermining climate science as true science, the CA accuses the environmental movement of being a "radical religion" bent on replacing Christianity (Beisner 2015a: para. 1). These sentiments were echoed a few times in my online conversations. For example, Michael accused environmentalists of "believ[ing] in a deity (Gaia) that will unleash her wrath if Earth is not returned [to] pre-industrial CO_2 levels" and Abigail[2] noted that "Global warming seems to be a secular religion for many." The CA (n.d.-e: 1:12–1:17) argues that environmentalists espouse a competing religion that names itself science to grasp at legitimacy that it does not actually have. Environmentalism "offers its own doctrines of God and creation" that directly challenge the Bible. An environmental-based religion is spread through the "siren call" (CA 2017: i) of alarmism that seeks to undermine the CA's work. The CA argues that environmentalists hide behind the label science to justify their own corrupted version of Christianity. But it is only with Christianity, and not the false religion of environmentalism, that societies can "flourish intellectually, morally, aesthetically, and materially" (CA n.d.-b: para. 8).

It may seem contradictory for separators to use calls of being a "religion" as an insult, because the accusation implies that religion is not a meaningful decision-making tool. But, this rhetorical move can be interpreted as non-contradictory if one considers the separators' organizing framework. In the separators' melodrama, there is only good and evil. Separators and true Christians are on the side of good, while others are on the side of evil, compromising Christian values and inverting God's hierarchy. Being on the side of evil, environmentalists cannot be Christian, so their religion must be a different, distinct one. Under this framework, separators call environmentalism a "religion" to ascribe the negative characteristics of a *false* religion, and not the positive characteristics of Christianity, to the movement. Prelli (1989: 91) called debates over what does and does not count as science a process of "demarcation." Demarcation is an argument about appropriate labels and rules for who can participate, negotiated anew in each interaction, which shifts the focus of the conversation onto debating those labels.

For the CA, only scientists who support Christian values and interpretations count as "real" or "true" scientists. These phrases function as rhetorical gatekeepers, whereby interlocuters must first overcome doubts that they are, in fact, scientists, instead of addressing more prominent concerns over what to do with the scientific information at hand. When I first contacted Beisner for an interview, he prompted me to fill out an extensive survey that asked highly technical questions about climate change. The questions spanned a variety of topics related to climate change including carbon emissions and rates of sea level rise. He advised me that no one should participate in climate discussions before they could answer all of the questions (without looking up any reference material). Not only did this request serve to disarm and surprise me, but it also gave the brief perception that I was interacting with someone who could answer those questions and thus was more scientifically knowledgeable than I was. This is a known strategy of the CA, as journalist Leo Hickman was asked by Beisner to read the CA's most recent book, *Resisting the Green Dragon: Dominion not Death* before agreeing to an interview (Hickman 2011). To argue against either request is to set oneself up to defend ignorance (of climate statistics or the book's content) and fall into the discursive trap of challenging definitions and ceding control of them to the CA.

These gatekeeping tactics serve as a way to control definitions and who gets to participate in discussions. If they can label mainstream climate science a religion, religious environmentalists traitors, and their own perspectives as scientific *and* proper Christian stewardship, then separators have won the rhetorical battle over framing the discussion. Winning this battle is not simply a game of words, but for separators, it is a fight for the presence and flourishing of Christian morals, ethics, and values. In seeking to restore "the fundamental Christian doctrines of God, creation, humanity, sin, and salvation," the CA (n.d.-f: para. 3) acts according to its guiding framework, views environmentalism as a threat, and undermines it whenever possible. In this sense, we can consider the CA and separator attitudes and behaviors as understandable and even logical, given their orienting worldview. But, that must not deter us from evaluating the fallout of controlling definitions as harmful to productive deliberation in the climate change controversy.

For example, when the CA summarized in their statement of faith that environmentalism "fails the tests of theology, science, and economics," they shift the conversation from one about deliberative action to a forensic one about labels, standards, and categorization (CA 2009b: 1). People wishing to engage with separators are then doubly tasked with countering the proposed labels *and* restoring the conversation back to its original frame of environmental action. If the goal is to engage separators on environmental topics, we first recognize the importance of labels and values to them and how easily one might become synonymous with enemy. As I will discuss more in the following chapter, my conversation with a particular separator quickly spiraled out of control because we could not agree on definitions, resulting in her calling my terms "ill-defined" and my attempts at continuing the conversation without proper terms a "futile endeavor."

Shifting blame

In addition to controlling definitions, separators engage in the shifting of blame for the present state of environmental discourse. Through this strategy, separators lay the burden of proof on environmentalists to counter the separators' war narrative, defend their status as scientists, and redefine stewardship. These topics of conversation serve as gatekeeping devices by which environmentalists and potential dialogue partners have to overcome differences over definitions, appropriate values, what "counts" as science or stewardship before real, productive dialogue can take place. The good people, the heroes of the narrative, must defend themselves against the perceived bias, manipulation, and secularism of the environmental movement. It is thus up to the environmentalist, scientist, or dialogue partner to disprove these assumptions and preconceived ideas instead of the separator justifying their position with evidence and rationality. Under their war framework, shifting the burden of proof is logical. In other words, why should the obvious hero and savior of humanity need to justify their position? Shifting the burden of proof is an argument strategy also used by the creationist movement, where creationist proponents do not provide evidence in support of their position, but simply find gaps or ask questions to undermine evolutionary perspectives (Pennock 2003). This rhetorical strategy can be quite effective, as the separator may appear to be engaging in conversation, but they are in fact, intentionally or unintentionally, deflecting true dialogue by unfairly putting the entire burden of proof on their dialogue partner. Although some scholars advocate that taking on the burden of proof is useful for establishing trust in climate conversations (Goodwin and Dahlstrom 2014), it is also important to acknowledge that accepting such a burden can create imbalance, which provides climate skeptics immense argumentative latitude.

By framing climate discourse as a warfare, the CA accuses environmentalism of fueling the war, an accusation that must be mitigated before more productive conversations can be had. For example, Beisner (2012: paras. 1–2) argued that people today are in the midst in "a spiritual world war" where Satan, "who hates God but cannot attack Him directly," works to "threaten the sanctity of human life; the dignity of human sexuality, marriage, and the family; and the God-given mandate for human beings to exercise godly dominion over the Earth" through "the Green movement." The CA interprets the controversy over climate change as "warfare" (Beisner 2010: para. 21) between religious and environmental worldviews over the impact of human activity on the environment and the appropriate response to it. The CA described the "debate over environmental stewardship" as "a *clash* of worldviews, with conflicting doctrines of God, creation, humanity, sin, and salvation" (n.d.-c: para. 2). The CA proudly states that it intervenes in this conflict in "a purposely adversarial way" to combat the violence of its enemies (Spencer 2017: para. 2).

These tactics of engagement include the publication of articles, videos, and blogs to compete with environmental messages in the mainstream media. For example, the CA has released a six-hour video series that features interviews with climate deniers and religious leaders. Titled, "Resisting the Green Dragon," the video series compares climate change to a dragon: powerful, terrifying, dangerous, but only a fantasy. On the video series' website, the CA justifies the naming of the series using James 4:7: "Submit therefore to God. Resist the devil and he will flee from you" (CA n.d.-e: para. 3). The environmental movement is characterized as both a dragon and the devil, who must be resisted and defeated. Because of the threat of the environmental movement, the CA urges that "people interested in truth need to learn to combat" its teachings and presence in society (n.d.-a: para. 7). In the battle against environmentalism, the CA predicts that its side will be the victor, as environmentalism is often "self-defeating" (2009a: 9).

The repetitive use of terms such as attacks, threats, defeat, adversaries, and con-flict drives a wedge between who the CA considers true stewards of the environ-ment and a brand of stewardship that has been corrupted by the environmental movement.

The CA's melodramatic frame forms a lens through which all actions by environmentalists are viewed as threatening attacks on the CA and its beliefs. The CA calls the environmental movement "the greatest threat" (Legates and van Kooten 2014: 1) to humanity, because of the ways that it will "destroy," "endanger," and "condemn" society (CA 2009a: 1). Environmentalism is dan-gerous because it will be "devastating to the world's poor," "threatens the sanc-tity of life," and is "targeting our youth" (CA n.d.-e: 0:33–0:34). The CA warns that "millions [are] falling prey" (CA n.d.-e: 0:33–0:34) to the evils of environ-mentalism. In one article, Beisner (2017f: para. 1) characterized environmental marches as violent acts of "thuggery." Climate science and environmentalism are viewed as enemies who interact with faith violently and aggressively, espe-cially toward the CA. The CA depicts itself as "the target of vicious attacks" that "are not only irrational but also morally disgusting" (Beisner 2017a: paras. 1, 21). The CA and environmentalism are thus "nemeses" who are "fighting on many battlefronts" over the proper actions including domination over public thought and policy action (Beisner 2012: para. 19). The CA refers to climate science as having committed "premeditated murder" in its commitment to the scientific consensus on global warming (Beisner 2010: para. 1). If the CA and Christianity are under attack, the burden falls to environmentalists to refuse those accusations and re-characterize their actions.

The CA attributes the corruption of the Church to the environmental movement, which "has in the last twenty years ... targeted the world's religious communities ... infiltrating them to 'Green' their messages" (Beisner 2012: para. 36). The CA equates the Creation Care movement with "an invasive, hybrid species that over-shadows the cross of Jesus Christ" (Beisner 2014: para. 1). Using an analogy of an invasive plant characterizes environmentalism as foreign and non-native to Chris-tian thought. Christian environmentalists are viewed as weeds that have been "watered and generously fertilized by Left-wing foundations" (Beisner 2014: para. 1). The invasive plant metaphor evokes the Biblical story of the snake in the Garden of Eden. Both the invasive plant and snake might look, on the surface, like they belong in the garden, but their presence is dangerous and corruptive. The CA has resisted environmental infiltration and stands by its commitments to "recover the blessings of Genesis" and "godly dominion over the Earth" (Beisner 2012: paras. 53, 55). One of my separator dialogue partners, William, echoed the CA's ideas of infiltration, arguing, "Individuals exist in every field, charitable and other-wise, who would use good causes to further evil agendas." Abigail further noted that when hearing environmental messages, we must think, "What's the incentive for the messenger?" Both of these statements imply that pro-climate messages are inherently biased and flawed, so all messengers must also be suspect. It is thus up to Christians and environmentalists to prove their interpretation is more Biblically accurate while deflecting accusations of having fallen to corrupting influences.

The CA does not have to defend its actions or approach to the climate change controversy; it is the behaviors of the environmental movement that have justified them. The burden of proof is placed on the environmental movement to characterize the controversy as something other than a war, to restore their reputation as scientists and not as criminals or aggressors, and keep the conversation going despite these mischaracterizations. If environmentalists are threatening Christian values, morals, and even people, then the CA and other separators are not only justified but morally obligated to defend their beliefs. Furthermore, they are justified in reacting as strongly as they perceive that they are being attacked. Shifting blame and thus the burden of proof onto environmentalists is a rhetorical strategy to change the focus of conversation to an arena where skeptics have the upper hand. Refusing to engage until these preliminary topics are addressed, such as definitions, stalls conversations and keeps environmentalists on the defensive. This also restricts the substance of the controversy from emerging, keeping the focus on the rules for engagement in the controversy or the "parameters for public discussion" (Olson and Goodnight 1994: 250). When engaging separators, these rhetorical features may emerge that make conversation difficult and potentially unmanageable, especially as attempts to change or challenge definitions may only reinforce the separators' skepticism and wariness of other perspectives.

Appealing to authority

In addition to framing the rules for conversation and the definitions being used for discussion, separators also limit relevant and appropriate authorities from which dialogue partners can draw. The Bible serves as the measure by which separators prioritize their values and from which they get their sense of moral truths. The CA argues that climate change should only be addressed "from [a] Biblical Worldview perspective" (Kinard 2015: para. 2). Separators thus perform a type of tautology in that the source of guidance for decision-making and their justification for following that guidance both rest in the Bible. The CA believes that the Bible is the only logical starting point, because it is "the sole, absolute, inerrant epistemological basis for mankind [*sic*] for all knowledge of all things, seen and unseen" (n.d.-b: para. 3). The CA paints mainstream science and environmentalists as enemies because they make "false and harmful" claims that "contradict" a literal interpretation of the Bible (CA n.d.-b: para. 3). Other starting points that dismiss or run counter to the Bible are thus, by the CA's definition, flawed. The CA (n.d.-b: para. 4) denies science's position that "the physical universe and human observations" contain truth or can serve as an appropriate authority when making environmental decisions. Beisner (2017b: para. 8) argued that the CA's environmental positions are based on "evidence" while environmentalists base their opinions on "climate scaremongering," which "play[s] lackey to politics." The CA perceives itself as having the weight of the Bible on its side and accuses environmentalists as being beholden to political predispositions and ideological commitments. The CA proclaims that environmentalism is not reasonable or rational. Instead, the CA calls it "speculative prognostication driven by a precommitment to fearsome scenarios" (Beisner 2017e: para. 15). The Bible is the only rational a priori commitment.

The CA (n.d.-f: para. 3) describes the environmental movement as being guided by a "worldview, theology, and ethics [that] are overwhelmingly anti-Christian" and that aim to "undermine the fundamental Christian doctrines of God, creation, humanity, sin, and salvation." For the CA, any worldview that starts outside of the Bible is considered contradictory to Christianity as a whole. The CA considers the ongoing "debate over environmentalism" to be not only about what people should do, if anything, about the environment, but also "about fundamental worldviews and their implications of ethics" (Beisner 2017d: para. 7). While some environmental groups define stewardship as protecting the environment, the CA sees the two as separate: "we must never conflate Biblical [E]arth stewardship with environmentalism" (Beisner 2015a: para. 2). The CA argues that "no other philosophy, religion, or worldviews provides a sufficient basis for stewardship of creation" (2009a: 9). The CA's stewardship is based on Biblical standards, while other brands of stewardship "deify nature, degrade people, or disregard the needs of the poor" (2009a: 9).

For the CA, humans are the exalted head of God's creation and thus have the right to use it for economic progress and survival. CA Blog contributor Vantassel (2016: para. 4) argued that when God created the Earth, He dictated that "some creatures are more valuable than others and those lower creatures are to serve the higher creatures." Following Western hierarchy of humans over non-human life (Milstein and Dickinson 2012), the CA reifies this hierarchy and considers it part of discipleship and honoring God's intended relationship between humans and other life. Inverting this hierarchy or lauding nature as equal to humanity is to worship the false God of the "green religion" (Taylor 2004: 992) and to undermine God's preferred stratification of humans as distinct from other life. The CA argues that when the prescribed hierarchies of the Bible are challenged, humans transform "from ruler to slave of the Earth" (Beisner 2012: para. 19), abandoning their post of dominion clearly stated in the Bible. It is thus the Bible's description of the proper order of things that becomes the foundational reference for the performance of stewardship. Other definitions, which emerge from different sources, are, by default, rejected.

Because the CA believes that God is in complete control over His creation, the group does not see the need to address environmental concerns. In its list of affirmations and denials, the CA rejects the idea that God "would have made [the Earth] susceptible to catastrophic degradation" and instead affirms that "Earth and its physical and biological systems are robust, resilient, and self-correcting" (CA n.d.-b: paras. 16, 17). This list pairs one affirmation and one denial on important subjects, which solidifies how the CA draws battle lines. One either agrees with these tenets or stands in opposition to them. Environmentalism and mainstream climate science, which largely disagree with the CA's affirmations and advocate for what the CA denies, sin against Christianity and participate in the "rejection of the Creator" (n.d.-b: para. 8). In a conversation I had with a separator, Isabella echoed the CA's belief in the resilience of the Earth by arguing that if "a day were to come where the end of the entire human race was at hand I do think God would interfere." Isabella disregarded the possibility for "a totally irreversible accidental annihilation" at our own hands. Instead, she argued that "if He wants us to roast ourselves alive then that's what will happen regardless of what we do." Isabella's interpretation of her faith drove her to consider human-caused disasters an impossibility; they were either preordained by God and thus not human-caused at all or God would intervene and stop anything from happening that is contrary to His plan. Because they view the Bible as the ultimate authority, mainstream climate discourse that contradicted or amended this perspective by forewarning about human-caused environmental catastrophe could either be dismissed as illogical or welcomed as a sign of God's ultimate plan (Barker and Bearce 2013).

In addition to separating themselves from mainstream scientific conclusions because of their incompatibility with the resiliency of the Earth, separators also do so because of the perceived negative effects of climate science and environmentalism, on culture and society. The CA warns that when Christianity is removed from the public sphere, its positive moral foundations will disappear as well. The war they fight is for the wellbeing of society and the reestablishment of Christianity, lest the whole nation fall to immorality and ruin. For the CA, appealing to the Bible as the ultimate authority is not only an issue of proper environmental decision-making, but is also an issue of society's health and wellbeing overall. Referring to the environmental movement, the CA (2009b: 1) argues:

> [The environmental movement] rests on and promotes a view of human beings as threats to Earth's flourishing rather than the bearers of God's image, crowned with glory and honor, and given a mandate to act as stewards over the Earth – filling, subduing, and ruling it for God's glory and mankind's benefit. It either wrongly assumes that the environment can flourish only if humanity forfeits economic advance and prosperity or ignores economic impacts altogether. And in its rush to impose draconian reductions in greenhouse gas emissions, it ignores the destructive impact of that policy on the world's poor.

When victims are threatened and need to be rescued, there are no sacrifices too great nor challenges too daunting. The separators are on a moral crusade, a holy mandate, to eradicate their enemy and restore morality to humanity and vulnerable populations. Melodramas justify these crusades by "amplifying [the] moral and emotional dimensions" of controversy (Schwarze 2006: 240). What may at first appear to be a disagreement about what we know and how we know it about the Earth and its climate is transformed into a controversy about the moral fiber of humanity. People and groups who make use of these guiding terms and frames thus rhetorically raise the stakes of environmental discourse to include much more than climate science, shifting the conversation away from areas in which environmentalists and scientists may be viewed as authorities.

The CA specifically focuses on the economic damage that environmentalism will do to the global poor. The CA equates environment advocacy with "regressive taxes – taxing the poor at higher rates than the rich," and placing the economic burden for adopting green technologies on developing countries (CA n.d.-d: para. 11). Because the environmental movement privileges the environment over human life, the CA argues that it will destabilize the economy and cripple businesses. The CA believes that vulnerable populations worldwide will be adversely affected by environmental regulations that inhibit growth out of poverty. In other words, those who will be most affected by the environmental movement are the inhabitants of developing countries that will be at the mercy of how industrialized society responds to climate change. The CA thus inverts the notion of climate justice, which is that those most vulnerable to the consequences of climate change have contributed the least to it (de Onís 2012). From the CA's perspective, environmental action cannot change the environment, because God has ultimate control over the Earth. But, on its foolhardy mission to try and save the environment, the CA argues that environmentalism will have devastating effects on the global economy. In "An Evangelical Declaration on Global Warming," the CA (2009a: para. 1) argued that current environmental policies will:

> destroy jobs and impose trillions of dollars in costs to achieve no net benefits. [Environmental policies] could be implemented only by enormous and dangerous expansion of government control over private life. Worst of all, by raising energy prices and hindering economic development, they would slow or stop the rise of the world's poor out of poverty and so condemn millions to premature death.

In addition to lauding the authority of the Bible, separators undermine competing authorities. For example, the above passages serve both to amplify the CA's claim to authority and to undermine the decision-making capacity of environmentalists. The CA's rhetoric paints a pessimistic and dystopian view of the future if environmentalists win the war. The CA's discourse may even be characterized as apocalyptic, the same claim conservatives often levy against environmental activists (Bloomfield and Lake 2015). Jennifer Peeples *et al.* (2014: 228) argued that to "identify environmentalism as apocalyptic" is to "mark environmentalism as radical, outside the mainstream, and unreasonable." The biased, hysterical environmentalists cannot be trusted to make decisions about the future of the Earth, which they themselves endanger. In the same breath of denouncing the scare tactics and "gloom and doom" (Whelchel 2017: 2) of environmentalism, the CA engages apocalyptic rhetoric as well. The CA's apocalypse, however, is not an environmental one, but an "industrial" one, that warns of risking the economy and the poor for the betterment of the environment (Peeples *et al.* 2014: 229).

The real threat, the true apocalypse that our world faces, does not come from environmental disaster but "economic catastrophe" (Peeples *et al.* 2014: 240).

The CA associates economic failure with an abandonment of the poor and those most in need. Similarly, the adoption of capitalism and a "free-market economic system" are examples of the "flourishing" that all will experience "in God's kingdom" (Whelchel 2017: 11). The CA draws its moral foundations from the Bible and argues that certain verses (i.e., Proverbs 14:31 and Proverbs 21:13)[3] necessitate the protection of the poor as utmost priorities. When society focuses on economic freedom instead of the environment, the CA argues that there are many benefits, including "higher incomes (even among the poor), greater happiness, better protected civil rights, cleaner environments, and longer life expectancy" (Whelchel 2017: 14). The CA implies that a focus on the economy will lead to environmental benefits naturally, without the need for direct environmental intervention. Environmentalists, therefore, would best achieve their goals by abandoning their green religion and following the CA's brand of stewardship. The CA thus encourages the practicing of true "creation stewardship out of love for God and love for our neighbors – especially the poor" (n.d.-d: para. 14) by following the Bible's authority and abandoning environmentalism.

Appealing to authority is a strategy that complements and, in some ways, undergirds the previous ones. When every answer is rooted in their particular interpretation of the Bible, their specific definitions, and their selected authority, separators construct all the terms and thus, the overall orientation to the controversy and conversation at hand. Appealing to authority is a strategy that legitimizes the CA's answers to questions about the environment, climate science, economics, and morality, and that roots their unshakable hierarchy in a text they consider immovable and eternal. If nothing can shake this authority, then the warrant that the Bible holds all of the answers and should be followed will always serve as valid and reasonable support for the separators' views on and approach to climate change and climate change communication.

Conclusion

This chapter highlighted the guiding terms of the separators and how those terms influence the rhetorical features that emerge in separators' discourse when engaging environmental topics. Taken together, these strategies align with the values, morals, and priorities afforded by a worldview dominated by notions of war and defense. Strategies of controlling definitions, shifting blame, and appealing to authority construct a strong defense against environmental arguments that give separators the upper hand in conversations. Together, they serve to preempt arguments rooted in the value of the Earth, liberal ideologies, and the authority of climate science. These rhetorical features may also shut a conversation down before it gets started, especially if a dialogue partner takes the bait, as it were, and rails against the provided definitions.

Employing war, melodrama, and morality as their orientating terminology, separators respond to the climate change controversy by separating environmentalism and Christianity on opposing sides. The separators loudly proclaim their side to be the heroes who defend values and morals against environmentalism.

The perceived negative moral, cultural, environmental, and economic consequences of climate science necessitate strong and powerful responses. The separators' rhetorical melodrama requires the strong opposition and potential sacrifice of scientific villains to reestablish the Christian order and its dominance in society. At stake for the CA is the future of humanity, its morality, and the flourishing of Christian values. The polarization inherent in melodrama creates a heightened sense of urgency and severity, motivating separators to avoid compromise and act swiftly and resolutely.

The next chapter will outline creative, subtle approaches to engage separators that aim to avoid disengagement and aggression that can emerge during environmental discourse. A consideration of the separators' rhetorical features, their terministic screen, and guiding metaphors is crucial for engaging separators in conversations about the environment and encouraging them to rethink and to reimagine interactions with environmentalists. We want to encourage separators to view conversations about the environment not as battles in the larger war of the climate change controversy but as opportunities for mutual understanding and productive engagement.

Notes

1   Then God said, "Let us make mankind in our image, in our likeness, so that they may rule over the fish in the sea and the birds in the sky, over the livestock and all the wild animals and over all the creatures that move along the ground."

(Genesis 1:26, New International Version)

2 We will see quotations from Abigail again in Chapter 3, because Abigail's discourse most prominently aligns with the worldview of the bargainers. Some of her arguments, however, were quintessential of separators, which is why I quote her here as well. Abigail, among others I interviewed, provide evidence that the categories in the separator-bargainer-harmonizer typology are fluid. While people will primarily fit in one of the categories, they may draw argumentative resources and strategies from multiple categories.
3 Proverbs 14:31, "Whoever oppresses the poor shows contempt for their Maker, but whoever is kind to the needy honors God." Proverbs 21:13, "Whoever shuts their ears to the cry of the poor will also cry out and not be answered."

References

Abigail [pseudonym]. (2018) personal communication [digital exchange].
Barker, D.C. and Bearce, D.H. (2013) End-Times Theology, the Shadow of the Future, and Public Resistance to Addressing Global Climate Change. *Political Research Quarterly* 66(2): 267–279. DOI: 10.1177/1065912912442243.
Beisner, E.C. (2010) Wanted for Premeditated Murder: How Post-Normal Science Stabbed Real Science in the Back on the Way to the Illusion of "Scientific Consensus" on Global Warming. Cornwall Alliance, March 18. Available at: https://cornwall alliance.org/2010/03/wanted-for-premeditated-murder-how-post-normal-science-stabbed-real-science-in-the-back-on-the-way-to-the-illusion-of-scientific-consensus-on-global-warming/ (accessed April 4, 2018).

Beisner, E.C. (2011) Huffington Post: Is Cornwall Alliance Responsible for Death of Environmentalism in America? Cornwall Alliance, May 31. Available at: https://cornwallalliance.org/2011/05/huffington-post-is-cornwall-alliance-responsible-for-death-of-environmentalism-in-america/ (accessed December 22, 2018).

Beisner, E.C. (2012) Putting Together the Pieces in the Spiritual World War. Cornwall Alliance, October 1. Available at: https://cornwallalliance.org/2012/10/putting-together-the-pieces-in-the-spiritual-world-war/ (accessed April 4, 2018).

Beisner, E.C. (2014) Evangelical Environmentalism: Bought and Paid for by Liberal Million$$$? Cornwall Alliance, October 31. Available at: https://cornwallalliance.org/2014/10/evangelical-environmentalism-bought-and-paid-for-by-liberal-million/ (accessed April 4, 2018).

Beisner, E.C. (2015a) The Radical Religion of Environmentalism. Cornwall Alliance, October 8. Available at: https://cornwallalliance.org/2015/10/the-radical-religion-of-environmentalism/ (accessed April 4, 2018).

Beisner, E.C. (2015b) World Magazine Exposes Evangelical Environmentalists' Growing Dependence on Green-Left Funding. Cornwall Alliance, May 30. Available at: https://cornwallalliance.org/2015/05/world-magazine-exposes-evangelical-environmentalists-growing-dependence-on-green-left-funding/ (accessed April 4, 2018).

Beisner, E.C. (2017a) Dear Media: We Don't Have to Agree to Have Intelligent, Friendly, Discourse. Cornwall Alliance, September 1. Available at: https://cornwallalliance.org/2017/09/dear-media-we-dont-have-to-agree-to-have-intelligent-friendly-discourse/ (accessed April 4, 2018).

Beisner, E.C. (2017b) Earth Day 2018 and the Anti-Scientific "March for Science." Cornwall Alliance, April 20. Available at https://cornwallalliance.org/2017/04/7459/ (accessed April 4, 2018).

Beisner, E.C. (2017c) EPW Chairman Introduces Cornwall Letter Supporting Pruitt EPA Nomination into Hearing Record. Cornwall Alliance, January 18. Available at: https://cornwallalliance.org/2017/01/epw-chairman-introduces-cornwall-letter-supporting-pruitt-epa-nomination-into-hearing-record/ (accessed April 27, 2018).

Beisner, E.C. (2017d) Heidegger, Fascism, Evergreen State College, and the Environmental Movement. Cornwall Alliance, June 19. Available at: https://cornwallalliance.org/2017/06/heidegger-fascism-evergreen-state-college-and-the-environmental-movement/ (accessed April 6, 2018).

Beisner, E.C. (2017e) The Ice Is Melting! The Ice Is Melting! Run for Your Lives! Cornwall Alliance, June 22. Available at: https://cornwallalliance.org/2017/06/the-ice-is-melting-the-ice-is-melting-run-for-your-lives/ (accessed April 4, 2018).

Beisner, E.C. (2017f) Peoples Climate Movement – Not Science, Just Thuggery. Cornwall Alliance, May 8. Available at: https://cornwallalliance.org/2017/05/peoples-climate-movement-not-science-just-thuggery/ (accessed April 4, 2018).

Bellah, R.N. (1991) *Beyond Belief: Essays on Religion in a Post-Traditionalist World.* Berkeley, CA: University of California Press.

BHAN, Satya (2017) The Love of Giving. YouTube video, 1:56, March 18. Available at: www.youtube.com/watch?time_continue=37&v=YNx7kDJ2kDI (accessed February 18, 2018).

Bloomfield, E.F. (2018) Argumentation in the Identity Politics of the Trans Selfie: Recovering Greek Mythology to Analyze Contemporary Gender Arguments. In: Lake, R.A. (ed.) *Recovering Argument.* New York; London: Routledge. Available at: www.routledge.com/Recovering-Argument/Lake/p/book/9781138294899 (accessed August 24, 2018).

Bloomfield, E.F. (forthcoming) Ecocultural Identity in the Creation Care Movement: Analyzing Contemporary Performance of Religious Environmentalism. In: Milstein, T. and Castro-Sotomayor, J. (eds.) *The Routledge Handbook of Ecocultural Identity*. New York; London: Routledge.

Bloomfield, E.F. and Lake, R.A. (2015) Negotiating the End of the World in Climate Change Rhetoric: Climate Skepticism, Science, and Arguments. In: Meisner, M.S., Sriskandarajah, N., and Depoe, S.P. (eds.) *Communication for the Commons: Revisiting Participation and the Environment*, pp. 384–396. Uppsala, Sweden: The International Environmental Communication Association.

Bloomfield, E.F. and Tillery, D. (2019) The Circulation of Climate Change Denial Online: Rhetorical and Networking Strategies on Facebook. *Environmental Communication* 13(1): 23–34. DOI: 10.1080/17524032.2018.1527378.

Burke, K. (1966) *Language as Symbolic Action: Essays on Life, Literature, and Method*. Berkeley, CA: University of California Press.

Burke, K. (1969) *A Grammar of Motives*. Berkeley, CA: University of California Press.

Burke, K. (1984) *Permanence and Change: An Anatomy of Purpose*. Berkeley, CA: University of California Press.

Cornwall Alliance (CA) (2009a) An Evangelical Declaration on Global Warming. May 1. Available at: https://cornwallalliance.org/2009/05/evangelical-declaration-on-global-warming/ (accessed December 1, 2018).

Cornwall Alliance (CA) (2009b) A Renewed Call to Truth, Prudence, and Protection of the Poor: An Evangelical Examination of the Theology, Science, and Economics of Global Warming. Available at: https://cornwallalliance.org/wp-content/uploads/2014/04/a-renewed-call-to-truth-prudence-and-protection-of-the-poor.pdf (accessed December 1, 2018).

Cornwall Alliance (CA) (2013) About the Cornwall Alliance. Available at: https://cornwallalliance.org/about/ (accessed December 1, 2018).

Cornwall Alliance (CA) (2015) Reasons to Sign An Open Letter on Climate Change to the American People and their Leaders: An Appeal to Christian Academics. November 18. Available at: https://cornwallalliance.org/2015/11/reasons-to-sign-an-open-letter-on-climate-change-to-the-American-people-and-their-leaders-an-appeal-to-christian-academics/ (accessed April 4, 2018).

Cornwall Alliance (CA) (2017) The Cornwall Stewardship Agenda. Available at: https://cornwallalliance.org/wp-content/uploads/2014/04/cornwall-stewardship-agenda.pdf (accessed December 1, 2018).

Cornwall Alliance (CA) (2018) Cornwall Alliance. Available at: https://cornwallalliance.org (accessed April 6, 2018).

Cornwall Alliance (CA) (n.d.-a) About Our Products. *Resisting the Green Dragon*. Available at: www.resistingthegreendragon.com/about-our-products-2/ (accessed April 4, 2018).

Cornwall Alliance (CA) (n.d.-b) The Biblical Perspective of Environmental Stewardship: Subduing and Ruling the Earth to the Glory of God and the Benefit of Our Neighbors. Available at: https://cornwallalliance.org/landmark-documents/the-biblical-perspective-of-environmental-stewardship-subduing-and-ruling-the-earth-to-the-glory-of-god-and-the-benefit-of-our-neighbors/ (accessed April 2, 2018).

Cornwall Alliance (CA) (n.d.-c) An Open Letter to Pope Francis on Climate Change. Available at: https://cornwallalliance.org/anopenlettertopopefrancisonclimatechange/ (accessed April 4, 2018).

Cornwall Alliance (CA) (n.d.-d) Protect the Poor: Ten Reasons to Oppose Harmful Climate Change Policies. Available at: https://cornwallalliance.org/landmark-documents/

protect-the-poor-ten-reasons-to-oppose-harmful-climate-change-policies/ (accessed April 6, 2018).

Cornwall Alliance (CA) (n.d.-e) *Resisting the Green Dragon: A Biblical Response to One of the Greatest Deceptions of Our Day* [DVD]. Available at: www.resistingthegreen dragon.com/ (accessed April 4, 2018).

Cornwall Alliance (CA) (n.d.-f) What We Do. Available at: https://cornwallalliance.org/ about/what-we-do/ (accessed December 29, 2014).

de Onís, K.M. (2012) "Looking Both Ways": Metaphor and the Rhetorical Alignment of Intersectional Climate Justice and Reproductive Justice Concerns. *Environmental Communication* 6(3): 308–327. DOI: 10.1080/17524032.2012.690092.

Desilet, G. and Appel, E.C. (2011) Choosing a Rhetoric of the Enemy: Kenneth Burke's Comic Frame, Warrantable Outrage, and the Problem of Scapegoating. *Rhetoric Society Quarterly* 41(4): 340–362. DOI: 10.1080/02773945.2011.596177.

Dixon, G.N. and Clarke, C.E. (2013) Heightening Uncertainty around Certain Science: Media Coverage, False Balance, and the Autism-Vaccine Controversy. *Science Communication* 35(3): 358–382. DOI: 10.1177/1075547012458290.

Evans, J.H. (2018) *Morals Not Knowledge: Recasting the Contemporary U.S. Conflict between Religion and Science*. Oakland, CA: University of California Press.

Goodwin, J. and Dahlstrom, M.F. (2014) Communication Strategies for Earning Trust in Climate Change Debates. *Wiley Interdisciplinary Reviews: Climate Change* 5(1): 151–160. DOI: 10.1002/wcc.262.

Hickman, L. (2011) The US Evangelicals Who Believe Environmentalism is a "Native Evil" Leo Hickman. *Guardian*, May 5. Available at: www.theguardian.com/environment/ blog/2011/may/05/evangelical-christian-environmentalism-green-dragon (accessed April 27, 2018).

Isabella [pseudonym]. (2018) personal communication [digital exchange].

Kinard, M. (2015) You're Invited! New Cornwall Documentary to Premier at the Heritage Foundation. Cornwall Alliance, October 2. Available at: https://cornwallalliance. org/2015/10/youre-invited-new-cornwall-documentary-to-premier-at-the-heritage- foundation/ (accessed April 6, 2018).

Konisky, D.M. (2018) The Greening of Christianity? A Study of Environmental Attitudes over Time. *Environmental Politics* 27(2): 267–291.

Lake, R.A. (1997) Argumentation and Self: The Enactment of Identity in *Dances With Wolves*. *Argumentation & Advocacy* 34(2): 66–89.

Lakoff, G. (1991) Metaphor and War: The Metaphor System Used to Justify War in the Gulf. *Peace Research* 23(2/3): 25–32.

Leavenworth, S. (2017) Trump Expected to Roll Back Obama Climate Initiatives Despite Easing of "Hoax" Rhetoric. *The Miami Herald*, January 21. Available at: www.miami herald.com/news/politics-government/article127944484.html (accessed April 27, 2018).

Legates, D.R. and van Kooten, G.C. (2014) A Call to Truth, Prudence, and Protection of the Poor 2014: The Case against Harmful Climate Policies Gets Stronger. Cornwall Alliance, September. Available at: https://cornwallalliance.org/wp-content/uploads/ 2014/09/A-Call-to-Truth-Prudence-and-Protection-of-the-Poor-2014-The-Case- Against-Harmful-Climate-Policies-Gets-Stronger.pdf (accessed December 1, 2018).

Liam [pseudonym]. (2018) personal communication [digital exchange].

McCammack, B. (2007) Hot Damned America: Evangelicalism and the Climate Change Policy Debate. *American Quarterly* 59(3): 645–668. DOI: 10.1353/aq.2007.0065.

McGee, M.C. (1990) Text, Context, and the Fragmentation of Contemporary Culture. *Western Journal of Speech Communication* 54(3): 274–289. DOI: 10.1080/10570319009374343.

Michael [pseudonym]. (2018) personal communication [digital exchange].

Milstein, T. and Dickinson, E. (2012) Gynocentric Greenwashing: The Discursive Gendering of Nature. *Communication, Culture & Critique* 5(4): 510–532.

New International Version. (n.d.) Available at: www.biblegateway.com/

O'Connor, B. (2017) How Fossil Fuel Money Made Climate Change Denial the Word of God. *Splinter News*, August 8. Available at: https://splinternews.com/how-fossil-fuel-money-made-climate-denial-the-word-of-g-1797466298 (accessed April 27, 2018).

Olivia [pseudonym]. (2018) personal communication [digital exchange].

Olson, K.M. and Goodnight, G.T. (1994) Entanglements of Consumption, Cruelty, Privacy, and Fashion: The Social Controversy over Fur. *Quarterly Journal of Speech* 80(3): 249–276. DOI: 10.1080/00335639409384072.

Peeples, J., Bsumek, P., Schwarze, S., and Schneider, J. (2014) Industrial Apocalyptic: Neoliberalism, Coal, and the Burlesque Frame. *Rhetoric & Public Affairs* 17(2): 227–253.

Pennock, R.T. (2003) Creationism and Intelligent Design. *Annual Review of Genomics and Human Genetics* 4(1): 143–163.

Prelli, L.J. (1989) The Rhetorical Construction of Scientific Ethos. *Evolution* 34(5): U980.

Ross, D.G. (2013) Common Topics and Commonplaces of Environmental Rhetoric. *Written Communication* 30(1): 91–131. DOI: 10.1177/0741088312465376.

Schwarze, S. (2006) Environmental Melodrama. *Quarterly Journal of Speech* 92(3): 239–261. DOI: 10.1080/00335630600938609.

Spencer, R.W. (2017) A Global Warming Red Team Warning: Do Not Strive for Consensus with the Blue Team. Cornwall Alliance, June 14. Available at: https://cornwallalliance.org/2017/06/a-global-warming-red-team-warning-do-not-strive-for-consensus-with-the-blue-team/ (accessed April 4, 2018).

Steinert, H. (2003) The Indispensable Metaphor of War: On Populist Politics and the Conrads of the State's Monopoly of Force. *Theoretical Criminology* 7(3): 265–291. DOI: 10.1177/13624806030073002.

Taylor, B. (2004) A Green Future for Religion? *Futures* 36(9): 991–1008. DOI: 10.1016/j.futures.2004.02.011.

Vantassel, S.M. (2016) Why Christians Don't Impose Veganism. Cornwall Alliance archive, January 12. Available at: www.archive.cornwallalliance.org/2016/01/12/why-christians-dont-impose-veganism/ (accessed April 6, 2018).

Whelchel, H. (2017) Economic Freedom and Care for the Environment: Mutually Exclusive or Mutually Beneficial? Cornwall Alliance, September 19. Available at: https://cornwallalliance.org/2017/09/economic-freedom-and-care-for-the-environment-mutually-exclusive-or-mutually-beneficial/ (accessed April 6, 2018).

William [pseudonym]. (2018) personal communication [digital exchange].

2 Separator strategies
Asking questions, accepting premises, and making it personal

The previous chapter outlined common rhetorical features of the category I call the "separators." People are categorized into this group when they are religious climate skeptics who turn to war frames, create distinct lines between friend and foe, and respond defensively to conversations about climate science and the environment. Due to the separators' frequent use of war frames, they are likely to put up the most resistance to engaging in dialogue on environmental topics. Whereas bargainers and harmonizers may be more open to conversation, mention of "the environment" or "climate change" may activate separators' defensive frames and cause them to presume manipulative or insidious intentions on the part of their conversation partner. Some separators may see the very broaching of these topics as a signal of pro-environmental attitudes, which would place their dialogue partner on the side of "enemy." Based on discourse analysis of climate skeptic groups and in-depth conversations with those who exhibit separator qualities (i.e., controlling definitions, using war metaphors, shifting burden onto environmentalists), I propose strategies to avoid activating those frames and to bring up environmental topics in less threatening and overt manners. Separators are largely committed to doubting climate science, distrusting authority figures, and fighting against environmentally-friendly policies. The goal, therefore, is not to "dislodge" their beliefs completely (Sangalang and Bloomfield 2018: 17), which is a tall order in general, let alone in a single conversation, but to start chipping away at established assumptions and to make new associations with environmental topics that align with separators' values.

The three strategies I discuss in this chapter are asking questions, accepting premises, and making it personal. These strategies are interrelated parts of an overarching strategy to listen to separators, acknowledge their beliefs, and incorporate separators' answers and values into the separators' own convincing. These strategies align with work by others who advocate giving of trust, committing to civility, and emphasizing listening (Goodwin and Dahlstrom 2014; Kirk 2018a; Ratcliffe 1999). Due to separators' war frame and skepticism of those perceived as enemies, the best person to convince a separator is themselves. By asking questions and accepting their starting positions, these beliefs and values form the premises of our counter-arguments. In this chapter, I outline how to implement these strategies in conversations with separators, including sample conversations where they have been used, explore existing theoretical and empirical support for these strategies, and describe approaches that should be avoided because they will likely close down conversation. Before discussing these best practices, I think it is important to address a particular type of separator whose attitude, demeanor, and perspectives will shut down conversation before it can begin.

A cautionary tale

Although I propose alternatives to thinking about skepticism on a continuum of severity, it is worth mentioning that the separators are the category most likely to contain extreme, polarized deniers. There are some people who will not waver or shift from their perspective. Furthermore, their agreement to engagement might be a rhetorical trap rather than a genuine attempt to listen and create mutual understanding. We might characterize these skeptics as "trolls" or "ideologues," who are not open-minded about environmental topics (Kirk 2018b). Dunlap (2013: 693) argued that there may be some deniers who have their "minds made up" and cannot be reached by any communicative intervention. Resolved skeptics could exhibit the characteristics of separators or bargainers, but their behavior is likely to be amplified under a war frame.

While I advocate that proper communication strategies can go a long way in opening up conversation and reaching deniers, even separators, it is important to recognize when conversation is a fruitless endeavor that may only cause our dialogue partners to double down on their skepticism of climate change (Hart and Nisbet 2012). As a cautionary tale before discussing separator strategies, I detail an encounter I had with an individual on an online forum who initially engaged me in conversation, but who quickly turned patronizing, aggressive, and accused me of being part of a conspiracy. In detailing this example of an unsuccessful conversation, I hope to help others identify potentially unproductive dialogue partners in their own interactions. These conversations also remind us that the strategies proposed here, though theoretically grounded and supported through practical engagement, might not always be successful or might not be met with civility from both sides.

After making a post on an online discussion website asking for people to reach out to me if they wanted to talk about their faith and the environment, Olivia commented that my questions and request for conversations were "vague and leading" and "a trap" to "shame/force others into government regulated environmental compliance." Although my post only contained a request for conversation and nothing more, Olivia immediately responded with assumptions about my own environmental beliefs and my intentions in asking for conversation partners. The mention of "the environment" in my question activated for Olivia the association that I must care about the environment to be asking questions about it, and if I cared about it, I was not on her side.

Paying attention to key words is a common feature of separators, who will point to certain terms as red flags of alarmism. These terms and their definitions played an important part of this particular skeptic's disengagement and hostility. Balking at terms perceived to be loaded is not unique in climate skeptical discourse. For example, conservative pundit Glenn Beck (2013) keeps a "keyword list" on the website for his book, *Agenda 21*, that he notes are code words environmentalists use when they talk about their plan to instill a totalitarian, one-world government. Included on that list are a series of words common to environmental conversations, such as "environment," "climate change," "Green House Gas," "emissions," "sustainable development," and "restoration" (Beck 2013). To use these terms in conversation is to risk identifying oneself immediately with those who separators believe are enemies. Perhaps even more concerning is the list's inclusion of seemingly unrelated terms such as "choice," "lifelong learning," "social justice," and "protest" (Beck 2013). A framework of war sets expectations for conversations and interactions with environmentalists, or simply those who disagree, as evil. These experiences reinforce the framework of war and establish stronger neural links between people's attitudes and behaviors (Lakoff 2010). For Olivia, the conversation was over before it began due to her established associations between environmentalism and deceit.

Attempting to be cordial and engage (as is the main thrust of my research), I attempted to continue conversation with Olivia. This, however, was a dead end, as she refused to engage further until I had answered a riddle about a tea party and subsequently ceased responding even after I had attempted to answer it. My interlocutor decided to end the conversation and I respected that decision. Conversations should not be forced upon others; they require the willing participation of multiple people. To treat an interaction as a matter of force is to view it unilaterally (Brockriede 1972) instead of thinking of it as a dialogue (Johannesen 1974). Of the utmost importance in engaging others is to enter a conversation committed to mutual understanding and respect, from everyone involved. Without those underlying values, the conversation will either be disengaged by one or both parties or will devolve into aggression and hostility. In other words, "There is no approach to persuasion that will not fail if an audience refuses to attend to a message" (Fisher 1985: 359).

Unfortunately, I am likely now an example in Olivia's mind of the evils of environmentalists, a risk that we necessarily undertake when engaging people on controversial topics such as climate change. What I viewed as a simple request was interpreted to be a threat to Olivia's worldview and Christian identity. I tell the story of this interaction with Olivia because it highlights how argument strategies, however rooted in theory and practice, can sometimes fail. Despite this cautionary tale, I had many pleasant conversations with people who exhibited separator qualities and found that some greatly enjoyed the opportunity to engage, discuss, and be heard. While we do not want to promote or validate climate skepticism, we do want to acknowledge the person behind those beliefs and the genuine emotions, values, and concerns that produce those beliefs. As Johannesen (1974: 96) argued, it is quite possible to "express judgment of policies and behaviors" a person holds while avoiding expressing "judgment of the intrinsic worth of [the] audience." The following sections outline strategies for engaging separators based on the patterns described in the previous chapter. After understanding that separators are guided by terministic screens of war and exhibit patterned rhetorical features due to that screen, it is important to avoid activating those connections and instead to emphasize a common desire for understanding.

Asking questions and accepting premises

The first two strategies of asking questions and accepting premises work together to serve a few purposes. First, they open separators up to share their own opinions and beliefs, which can help to create a welcoming, open, respectful environment. Through a more open approach and avoiding war metaphors, we can change the tone of the conversation from one of aggression and coercion to understanding and exchange (Brockriede 1972). The strategy of asking questions comes directly from our theoretical focus on enthymemes, which involve the audience in their own persuasion. Separators view themselves and their religious interpretations as the ultimate yardstick of truth and authority. Instead of combating that deep-seated belief, we can embrace it as a feature of separators' discourse and use it to frame our responses. Asking questions with the genuine commitment to listening can break down separators' instinct to approach conversations on the defensive and to view their conversation partners as on the attack.

One of my first conversations with a separator who I found through an online discussion forum started quite slowly. Benjamin was at first hesitant to engage me, as he noted that his opinion "deviates" from other responses that had already been posted that were more favorable toward the integration of faith and environmental attitudes. Taking their initial post as a sign of interest in discussing, but reluctance to express a minority opinion, I assured him that I was interested in hearing from a variety of different perspectives on how people integrate faith and the environment. With this reassurance, we soon scheduled a phone call. By not dismissing this person out of hand or undermining their skeptical perspective, I was able to gain their trust enough to engage in further dialogue (Goodwin and Dahlstrom 2014). During our call and after introducing ourselves, I asked a series of questions that started general (e.g., what do you think about nature and the outdoors or what does your faith say about the environment?) and then built from those standard questions based on answers to learn about their various views on the environment. I endeavored to learn what specific, individual views this separator held, so focused primarily on listening rather than talking or explaining.

When addressing all our categories, but especially separators, it is important to engage an audience-centered view of persuasion. This perspective first determines "how the audience thinks and acts" and then engages that information with the topic to which the speaker wants to persuade (Long 1983: 107). Based on Chaim Perelman and Lucie Olbrechts-Tyteca's (1969) *The New Rhetoric*, Long (1983: 107) argues that an audience-approach "fills the audience's consciousness with its very being," creating a "communion" with the audience to work as partners rather than adversaries. This approach to argumentation brings the audience into its own persuasion; the audience convinces itself of the speaker's message. If separators can be convinced that we are on the same team, that we both strive for a better future and the understanding of each other, then true dialogue, and not monologue (Johannesen 1974), can occur, which can be catalysts for self-reflection and discovery. Burke (1969: 544) argued that people must "identify" with one another before persuasion can begin. Once someone is convinced that "their interests are joined" and their beliefs are aligned, people can forge common ground, create trust, and come to understanding (Burke 1969: 544). In our communicative interactions, we must forge genuine connections with the people we engage and understand the person behind the skepticism.

In a conversation with Charles, he mentioned the uselessness of caring for the environment and animals beyond their "innate value as tools," as a "food source," and as "entertainment." These terms signaled for me that Charles's faith guided him to see the Earth, animals, and nature as irrelevant beyond human use. Charles thought that we should care for the environment only so that environmental degradation does not "become a problem for us," such as disrupting our food supply. This perspective was directly rooted in his interpretations of scripture. For example, Charles quoted Ecclesiastes 1[1] and noted that anything not in direct worship and service of God is "vanity" and thus a "meaningless" endeavor. He continued by noting that human attempts "to fix the world is pride." People's attempts to regulate "CO_2 emissions" and protect endangered species can be viewed as important to some humans but not to God. During this conversation, I learned that Charles believed in the apocalypse, which made it difficult for him to see the Earth as an integral component of human survival. This can be partially attributed to the fact that Charles's eschatology, or his views on the apocalypse, caused him to see the Earth as temporary and dispensable.

I prompted Charles to tell me more about his eschatology to learn more about his specific perspective on the environment. Charles explained that he believed in the Second Coming, which included the eventual return of Jesus, who does not require human help or assistance to "purge the evil" from the world. Only through Jesus, he argued, could the world be affected or changed. Charles felt that while it may seem that humans are impacting the environment, people would never be able to steer it off course from God's plan. This belief is quite common in evangelical eschatology, which often includes the notion that the rapture will occur "before the going gets too tough" (Barker and Bearce 2013: 269). At this point in the conversation, I recognized the values that Charles held in maintaining human life and his firm eschatological beliefs. If he had already connected reducing CO_2 emissions and protecting animal life (that serves no purpose to humans) with vanity and meaninglessness before the apocalypse, that would be a difficult association to disrupt and change. In other words, these were potential dead ends and terms that might trigger defensive frames. Instead of addressing the hierarchy of God–humans–nature, I accepted Charles's premise that the apocalypse will happen by prompting this separator further about whether he thought Jesus would return "in our lifetime."

The relationship between Christianity and eschatology is complicated, as there are a variety of different interpretations of the Christian "End Times" and appropriate actions that should be taken in response. For example, some Christians welcome climate change as a sign of the impending apocalypse and Second Coming of Jesus Christ (Barker and Bearce 2013), but others use the apocalypse as an urgent call to protect the environment (Bloomfield forthcoming). After Charles told me that he thought Jesus would not return in our lifetime or even soon, (noting that he was "not that lucky"), I asked if he was concerned that "there would be a space of time between now and when Jesus returns when the environment would be so damaged as to lead to human suffering." Charles affirmed this possibility, which seemed to give him pause to consider how human life might be damaged by ignoring the environment. In asking questions and accepting the answers from my dialogue partner, I sought to disrupt the God–human–nature hierarchy and to link the quality and survival of human life with the environment as can be seen in the following argument model.

Argument to encourage protecting the Earth

Premise 1: The environment is valuable only for use by humans.
Premise 2: The degradation of the environment will make it useless to humans.
Conclusion: The environment should be protected so it can continue to be useful for humans.

After considering how environmental damage could lead to human suffering, Charles repeated their hierarchy that human life is more important than other forms of life. It seemed that he found my considerations reasonable and was at least thinking about how environmentally-friendly attitudes were not necessarily incongruent with his ideas. I interpreted his need to reiterate his previous point as a sign that he was revisiting the God–human–nature hierarchy to see if it aligned with my proposed environmental attitudes. By accepting Charles's premises that human life is most important and the apocalypse is coming, I created an argument that did not directly counter or refute his values and beliefs. Instead, my asking of questions spurred different connections that may not have been made previously. If Charles had felt the apocalypse would happen in our lifetime, as did others I chatted with, I would have turned toward existing evidence of humans suffering due to climate change, such as the increased spread of disease, disruptions to food supply, and more severe weather patterns (e.g., Shuman 2010; US Global Change Research Program n.d.).

Without asking these questions and learning about my dialogue partner, I may not have succeeded in creating a convincing enthymeme for them to fill in their value hierarchy of humans over other life. For example, if I learned that my dialogue partner thought that the apocalypse would happen soon, arguments about the length of human suffering would likely have seemed unreasonable. Additionally, arguments that rely on warrants that disrupt the God–human–nature hierarchy might similarly have been rejected. We can compare these approaches in their Toulmin models.

Non-tailored argument

Claim: We should prevent environmental degradation.
Grounds: Because the environment, nature, and animal life will suffer.
Warrant: Nature and the environment are important to protect (could be read as nature and the environment are as important as or more important than human life).

Tailored argument

Claim: We should prevent environmental degradation.
Grounds: Because humans may suffer if the environment is too damaged before the apocalypse.
Warrant: We should do what we can to minimize human suffering.

Asking questions and accepting premises encourage our dialogue partners to reveal the most important, salient topics on the environment to them and what they value, which tells us something unique about their perspectives. We may also gain their trust in sharing our interest in listening to their perspective. As was the case with my conversation with Olivia, many separators are hesitant to engage on environmental topics and may even react aggressively because of their preconceived notions about environmentalists. Our mutual respect, trust, and discussion of guiding beliefs and values can then be leveraged into rethinking, expanding, or at least questioning previously held beliefs. Separators' instincts may be to reject environmental topics out of hand, so it is important to reframe conversations not as direct confrontations, but as opportunities to learn. If we enter conversations as persuasive, coercive encounters, we will likely put separators on the defensive or cause them to end the conversation. While it is important to identify someone's overarching frames and category, it is also important to treat individuals as individuals and explore their discourse and beliefs for nuances and unique markers that could help adapt messaging to be most successful. The next strategy for conversing with separators embraces this ideal and invokes the local, personal, and specific to engage separators with what they care about most. This strategy builds upon the previous ones by making inventional use of the topics discussed by the separators to locate points of persuasion.

Making it personal

Making it personal is a broad strategy that focuses on bringing the environment back to values, ideas, and beliefs that the separators holds. After learning more about the separator, conversation can build upon those existing foundations. We can thus avoid entering the conversation with potentially unwelcome information. By making climate change intimately personal and meaningful to the separator, they can more clearly see themselves as impacted by the issue. Scholars have found that tailoring message content to the audience can produce increased attention to the message and persuasive outcomes. We can use the first part of the conversation (asking questions to gather information) to find tailoring opportunities. Some aspects to look for or to focus on are their personal life experiences (Fisher 1984), current location or places lived (Dillow and Weber 2016), value hierarchies (Perelman 1955), and "objects of care" (Wang *et al.* 2018).

In terms of tailoring messages to people's life experiences, Fisher (1984: 8) argued that people find certain narratives (and what we might describe as narrative arguments) particularly compelling and reasonable when they "ring true" with our perceptions of reality. He called this idea, "narrative fidelity," where how faithful a story is to someone's reality will predict their likelihood to accept or entertain that narrative as rational and true (Fisher 1984: 8). When we accept the experiences that people have already gone through, we can absorb those specifics into our conversation, mirroring their worldview in our discourse. For example, a cherished childhood memory about National Parks could be leveraged as a warrant (i.e., National Parks are important to protect) to justify environmentally-oriented claims. This relates to the second in-roads of geographic location. Dillow and Weber (2016) found that by simply decreasing the geographic scope of an organ donation message from the nation to participants' state increased organ donation registration. We are wired to care about our personal experiences; they resonate with us. Limiting the scope of discussion to places that are meaningful to dialogue partners or are places they have physically been can help shrink gaps between the vague, foreignness of climate change and people's everyday lives. It is hard to get people to see and understand the dangers of climate change because there are many issues and concerns plaguing them in the present. When there are many other things to distract people, it is imperative to make the information meaningful and immediate to capture people's attention and concern. For example, Burke (1984: 37) argued, "A philosopher, if he has a toothache, is more likely to be interested in dentistry than in mathematical symbolism." While both may be important, the other seems more urgent and relevant to one's current condition.

We can also tailor our messages to emphasize the values that people hold. Perelman (1955) noted that there are certain arguments that cannot be won by rationality alone; values must be taken into consideration. Even if people can agree that a value is important, they may weigh them differently, engaging disparate "value hierarchies." Thinking about value hierarchies can help us find what our dialogue partners care about most and recognize that many climate skeptics may still care about the environment but may rank it as less important than their faith, personal autonomy, family's wellbeing, and other factors. Value hierarchies connect to Wang *et al.*'s (2018: 26) concept of "objects of care." They define objects of care as "connectors" that help to make climate change "personally relevant for the individual." For example, Wang *et al.* (2018) argued that appealing to youth and the wellbeing of children can leverage the attention of parents to feel a connection to environmental policies. Centering discussions on common values and topics people care about can help redirect the defensiveness that separators feel into care and sympathy.

While every separator will be different, we can expect certain patterns of values, priorities, and beliefs based on prior research and the examples collected in this book. Ross (2013: 92) agrees with this strategy, noting that identifying common topics is "vital" for presenting "environment-related information for differing stakeholder audiences." If we know what is valuable, important, and resonates with particular audiences, we can tailor our messages to meet audiences where they are. In my conversations with separators, they made frequent mention of freedoms (e.g., freedom of speech and religious expression), local communities, and personal autonomy as prominent sources of concern to reject environmentalism. In this sense, separators share characteristics with conservative ideologies, which are often focused on the autonomy of the self and are linked to neoliberalism (Asen 2017). Bloomfield and Sangalang (2014) argued that conservatives operate under worldviews that emphasize the individual agent, where personal responsibility and the freedom to act autonomously are emphasized in the narratives they tell. For example, conservative politicians may shirk the need for societal or scenic policies in favor of individuals taking care of themselves without welfare or government intervention (Bloomfield and Sangalang 2014; Bloomfield and Tscholl 2018). Quoting Brock (1965), Klumpp and Hollihan (1979: 10) argued that "conservatives focus their enactments of events on individuals, the protagonists of the socio-drama, while the more leftist interpretations draw the circle of responsibility more broadly."

The strategy of making it personal thus combines generalized patterns of what we might expect separators to value with specific information we learn from conversations about the individual within our various group memberships. Because separators doubt climate change and environmental protection as important, arguments appealing to the environment as a warrant are likely to fail. Although not using the terms of the Toulmin model, Elliott (2014: 244) argues that science communicators can use "indirect arguments" to convince people about environmental topics without directly invoking the environment as a priority. I would describe this strategy as maintaining a consistent claim (e.g., we should invest money in eco-friendly technology research), but switching the grounds provided to activate a more widely accepted warrant and worldview (Bloomfield and Tscholl 2018).

Direct argument

Claim: We should invest more money in eco-friendly technology research.
Grounds: Because it will help the environment.
Warrant: It is important/valuable to society to help the environment.

In this example, the claim might be immediately rejected by a separator because of the grounds made to support it. Conversely, taking Elliott's (2014) advice and taking into consideration the power of enthymemes, we could transform the same argument model by adjusting the grounds and the resulting logical link.

Indirect argument

Claim: We should invest more money in eco-friendly technology research.
Grounds: Because it will help spur the economy by creating new jobs in technology sectors.
Warrant: It is important to create jobs and spur the economy.

There are many reasons available to lend support to claims of enacting eco-friendly policies. By using indirect arguments, we can "appeal to other beneficial effects associated with the policies under consideration" without directly using environmental *topoi* (Elliott 2014: 244). If one takes into consideration that separators are likely to value a strong economy and personal autonomy, we can turn to economic or ideological grounds to support environmental claims. Even better would be to incorporate the separators' specific perspectives and beliefs into those grounds, making links and connections between the individual dialogue partner and the environment. This approach combines our strategies of asking questions, accepting premises, and making it personal.

During my conversation with Benjamin, I specifically prompted him to think about definitions, one of the guiding rhetorical features of the separators. In doing so, I endeavored to figure out what makes a Christian "Christian" for him, without assuming his opinion. Benjamin answered that true Christians value human life and evangelization and that while "environmental care is good," we cannot let it distract from doing "as the Lord commanded." I asked Benjamin if he felt that Christians who are environmentalists are misinterpreting scripture or are not real Christians. The separator responded that "advocating for the improvement of [the] environment" might be other Christians' way of showing "love to their neighbor," which he noted is "right in a certain perspective." When I asked if true Christians can incorporate advocacy for the Earth in their faith, Benjamin responded that this would be a good Christian thing to do as long as they do not neglect "spreading the [Gospel]" while protecting the environment. He concluded that it is not problematic for "us [C]hristians to care for our common household" as long as they prioritize "faith and moral issue[s]" over environmental protection. For Benjamin, caring for the environment was not a negative behavior; it only became so when care for the environment displaced other concerns.

As I continued the conversation, I learned quickly that it would be highly unlikely for Benjamin to support CO_2 regulations, because he believed that these policies "cut off small companies" and "hurt competition," which are economic concerns common in skeptical discourse (Tillery 2018). Although CO_2 emissions were off the table that did not mean that all potential solutions were. I asked Benjamin about his views on technological innovation and increasing funding for researching sustainable practices, because such research would both enhance business productivity and help the environment. I specifically prompted him to consider if he would support the government providing more funds for researching eco-friendly technology. Benjamin responded that they think government funding for technology is a good thing, because "technology improves our lives." By turning to technology, I had substituted a middle ground between the economy and the environment that satisfied Benjamin's economic concerns while finding alignment between his beliefs and pro-environmental policies.

After considering my point, Benjamin noted that he did not think environmentalists supported technology funding but instead more often supported "business regulations." I acknowledged that many environmentalists do advocate for business regulations, but just as skeptics are not homogenous, there are also a variety of beliefs that environmentalists hold. For example, I referenced the Environmental Defense Fund that has an arm devoted to promoting new technologies in support of "clean energy." Benjamin seemed intrigued by this information and wanted to do more research about it. In this conversation, there was evidence that Benjamin was considering that environmentalists may share some of his viewpoints (shrinking the gap between friend and foe) and that there was more research to be done about previously held beliefs. By disrupting an "illusory correlation" he had created between environmentalists and a commitment to business regulations, Benjamin's opinion on at least some aspects of environmentalist attitudes seemed to be slightly moderated (McFadden 2016: 3). McFadden (2016: 12) suggests that "effort[s] to decrease illusory correlations may be a more effective form of scientific communication" than taking an exclusive informative or explanatory approach. I consider illusory correlations to be warrants; by breaking apart these correlations and warrants, people will not have as clear of an established link between their grounds and claims. McFadden's (2016) arguments are in line with my rhetorical critiques that offer enthymemes and self-convincing as a powerful tool to combat science denial.

Through asking questions, using the information and values provided to me across these conversations, I worked from rejecting the environment as important and valuable, to seeing environmental protection as a way of practicing Christianity or at least as being in alignment with some of its tenets. Instead of focusing on climate change directly or combating beliefs about CO_2 emissions and God–human–nature hierarchies, I asked questions, found out more information, and connected existing parts of their faith to the environment. If the goal is to promote environmentally-friendly attitudes and reduce hostility toward environmentalism, we should follow the path of least resistance. In these conversations, I did not convince a separator to believe that climate change is happening, nor that climate science and environmentalism (which Lucas called "secular ideologies") are not threats to Christianity. But, I did engage separators in extended conversations, acknowledge their beliefs, and offered ways that they could explore how being a good Christian and environmental protection might go hand-in-hand. In a sense, these separators were moved not to change their perspective in its entirety but to destigmatize environmental protection and to challenge their framework that environmental topics should put them on the defensive.

Because of separators' warring framework, it is unlikely that their beliefs will shift completely, especially in one conversation. These conversations, however, provide some evidence that separators are open to and can find room for ecocentric attitudes under their worldview. Asking questions, offering premises, and making it personal are strategies that encourage separators to come to conclusions themselves while not challenging or undermining their beliefs. While the strategies I propose here may not convert separators to harmonizers, they may prompt more environmentally-friendly behaviors and attitudes, which can be seen as dampening climate change denial. Even with small changes, reconsiderations, and reframings, my hope is that adjustments to underlying frameworks and associations could grow into larger long-term changes.

Strategies to avoid

Based on the separators' rhetorical strategies, we cannot only predict successful argument responses, but also ones likely to backfire, to activate the separators' defensive frames, and to end the conversation prematurely. Perhaps one of the most obvious topics to avoid is the dismissal of the role of faith in the conversation. Because separators engage the environment from a religious perspective, it is imperative to acknowledge religious-based arguments as part of the conversation. Being perceived as insulting or patronizing about a separators' faith is likely to shut down the conversation or confirm suspicions that their interlocutor is an "enemy." If you are engaging with a separator and you identify as a Christian, this may seem to be a tempting route to follow. But, because of the separators fierce loyalty to their definition of certain terms, attempts at creating consubstantiality or "shared substance" (Burke 1969) with them may backfire. If a separator feels that someone falls outside their scope of what it means to be "Christian" or what proper "stewardship" is, then making arguments using those terms (but differently defined) will likely fail. To be viewed by a separator as a non-Christian, while professing to be a Christian, could perhaps activate defensive frames more so than non-Christians.

Another strategy to avoid is overrelying on scientific authority. Although appealing to authority seems logical given the consensus of scientific authorities on climate change, it is a potentially risky move when dealing with separators (and others who do not share trust in climate scientists as authorities). Wilchelns (1972) argued that appeals to authority and credibility incorporate both the speaker's personality and their public character. Public opinion, therefore, may influence a person's ability to be an authoritative source of information. While some may see climate scientists as authority figures, others reject them out of hand. Scholars have argued that the credibility of speakers in part determines how strong their appeals to logic and reasoning will be perceived to be (Segal

1991). Once a view on someone's credibility has been established, it may lie "beyond the reach of *logos*" or logical reasoning (Chamberlain 1984: 102). In other words, one's credibility may not easily be altered by appeals to reasoning or examples to the contrary.

Previous studies have found that an overreliance on ethos can be detrimental to an argument. In one study, when given a weak argument from a highly credible source, participants rated the argument far weaker than when the same argument was presented by a non-credible source (Tormala *et al.* 2006). Furthermore, Ceccarelli (2011: 217) warned that engaging with skeptics on matters of authority can unwittingly shift the conversation to a defense of climate science (and science in general) that can easily "devolve into elitist rants" and "dueling ad hominem attacks." Considering that there are other strategies to engage separators, it seems logical to avoid engaging climate scientists as authorities, at least early on in the conversation, and shift topics to other reasons to engage environmentally-friendly attitudes and behaviors.

When appealing to authorities in conversations with separators, the best strategy would be to wait for them to offer someone they find credible instead of offering one without prior validation. Considering that the establishment of credibility is fluid and subject to change, we cannot rely on certain authority figures being consistent argumentative resources. If we consider the establishment of ethos to be in part due to the relationship between speaker and audience (Rosenthal 1966), we can acknowledge that appeals to ethos will always be context-specific and variable. Below is an example of a failed strategy of appealing to authority in a conversation with a separator when I assumed there would a clear and easy route to persuade through ethos.

During my conversation with Benjamin, I asked how he defined being a good Christian. After discovering that he identified as Catholic, I immediately turned to Pope Francis as a potential connection between their faith and the environment (Burton 2014). Pope Francis has been a prominent voice in the environmental movement, with his encyclical *Laudato Si'* being a call to action for Catholics worldwide to think of the Earth as a sibling worthy of care. When I asked about his thoughts on Pope Francis and the environment, the tone of the conversation changed from mostly positive and forthright to slightly guarded. Benjamin responded that he heavily disliked the work that Pope Francis was doing and noted that "He is putting too much effort on questions that shouldn't be a priority of the Church and neglecting his real duty as the Vicar of Christ." Benjamin perceived Pope Francis as valuing the environment over other Christian priorities and thus viewed him as untrustworthy.

Even though Pope Francis is the head of Benjamin's faith, he argued that Pope Francis is "wrong" to be "focus[ing] on mundane issues" such as the environment while the evangelizing duties of the faith are suffering. He specifically spoke about the "Dubia," which was a series of questions that concerned cardinals asked Pope Francis when he was appointed Pope (Allen Jr. 2017). The Dubia itself and then the lack of a response from the Pope, led Benjamin to doubt the Pope's commitment to "spreading the gospel and defending the faith."

Despite how tempting it is use Pope Francis as an authority, it was fruitless in this instance, because in Benjamin's eyes, the Pope had violated his "duties as Vicar of Christ." Disregarding what this separator saw as an important duty for the office of the Pope undermined their faith in Pope Francis as a legitimate authority over certain matters.

When I asked if Benjamin felt conflicted about doubting the Pope and not following his views on the environment, he responded, "I don't. I exercise my right to disagree with the Pope in matters that doesn't [*sic*] concern the faith." Due to Benjamin's distinct definition of what "counts" as Christian and what does not, he could easily dismiss Pope Francis's musings on the environment as outside the appropriate realm of spiritual guidance. In an attempt to salvage the conversation and learn more about their environmental beliefs, I asked if Benjamin had found anything in the *Laudato Si'* that resonated with him. At this moment, Benjamin stated that he had "tried to read his encyclical" but did not get very far because "it was too long and boring." He continued, "I don't think it would change my view on his works as Pope." I asked if he would consider trying again to read it, as certainly the words of the Pope, even if people disagree, are important doctrine for Catholics to consider. At this point, I offered my own interpretation of the encyclical that the document makes a compelling case for the productive relationship between Christianity and the environment. I believe because we had been talking for a while (nearly an hour) and had developed a rapport, Benjamin agreed to my request to "try to read it again" and reach out with his "thoughts." When we ended the conversation, he agreed our exchange was enjoyable and it caused him to "refine my views" on the environment enough look at *Laudato Si'* again.

This conversation was almost derailed by appealing to what I had assumed was a shared authority. Because this authority did not meet the separators' definition of an authority, it fell flat as a counter-argument to his anti-environmental views. Indeed, it nearly ended the conversation and risked the mutual respect we had developed for one another. The lesson here is that invoking authorities is a potentially risky strategy. Even the Pope can be a victim of the melodramatic frames of the separators' worldview. To avoid appealing to untrusted authorities, we can use our same strategy of asking questions and listening to discern who might be "safe" authority figures and which might have already been labeled untrustworthy and lacking credibility.

Conclusion

The "separators" is a term I propose that represents a collection of patterned perspectives toward climate change from a religious perspective. The response strategies I have outlined here are meant to serve as guidelines for engaging climate deniers who see the controversy over climate change through a terministic screen of war and turn to the Bible as an ultimate authority. I have also outlined strategies to avoid because separators may prematurely end conversations. The primary goal in engaging separators should be to avoid activating war frames that can derail conversations and to instead acknowledge their perspectives, engage in a dialogue, and ask them about their values and underlying motivations. These strategies should avoid the more aggressive, polarized tendencies of the separators and open up a comfortable, welcoming space for separators to be heard. A frequent motif in the discourse of climate skeptics is that they are not being heard or respected for their beliefs. For people who feel that their religion, identity, and values are being ignored or even threatened, traditional persuasive approaches are likely to fail.

Climate change is not simply a matter of weighing evidence; for the separators, there is far more at stake. Acknowledging and respecting these values and beliefs is not only a standard practice of ethical communicators but is also a theoretically and practically-supported strategy to engage people in difficult and controversial conversations (Ratcliffe 1999). As described in the conversation with the resolved skeptic, there are potential risks involved when engaging separators. They may not want to engage, may become aggressive or insulting, or may seek to undermine the motivations and values of their dialogue partner.

My proposed strategies of asking questions, offering premises, and making it personal will not instantly convert separators away from climate skepticism. Although theory and my experiences provide evidence that they are productive in-roads, every conversation and every separator will be different, requiring tailoring and adjustments. It is my hope that these strategies, which stem from the separators guiding frames and subsequent rhetorical characteristics, will at the very least work to keep conversations going, maintain cordial relationships, and prompt internal reflection. By focusing on the individual in the conversation, we can displace separators from their tendency to aggrandize discussions of climate change into crusades over morality and Christianity. Keeping the conversation at the personal level, connecting with their values, and learning about their beliefs can help us shift separators' traditional assumptions about climate science and environmentalists to more productive avenues.

This chapter also sets the stage for viewing the differences between the separators and bargainers and thus the necessity for tailored approaches to climate communication targeting climate skeptics. The next chapter describes the second category I consider to be climate skeptics, which I call the bargainers. Separators and bargainers challenge climate change science from religious perspectives, but they do so by using disparate discursive tools. These differences between the groups' rhetorical strategies, worldviews, and values call for us to reconsider what might be successful responses to each group. Before exploring the differences between my recommendations for engaging separators and bargainers, it is imperative to examine the underlying frames and vocabularies that distinguish bargainers as a unique category of climate skeptics.

Note

1 "Vanity of vanities, says the Preacher, vanity of vanities! All is vanity" (English Standard Version). In the New International Version, "all is vanity," is translated as

"everything is meaningless." The Book of Ecclesiastes focuses on themes of meaninglessness and the void, arguing that the only way to true happiness and purpose is through God. These verses resonated with Benjamin as the correct path forward, resulting in their belief that environmental activism was a deviation from the right path and, thus, a fool's errand.

References

Allen Jr., J.L. (2017) Why Was Pope Francis So Quick to Answer These "Dubia"? *Crux*, October 23. Available at: https://cruxnow.com/news-analysis/2017/10/23/pope-francis-quick-answer-dubia/ (accessed March 3, 2018).

Asen, R. (2017) Neoliberalism, the Public Sphere, and a Public Good. *Quarterly Journal of Speech* 103(4): 329–349.

Barker, D.C. and Bearce, D.H. (2013) End-Times Theology, the Shadow of the Future, and Public Resistance to Addressing Global Climate Change. *Political Research Quarterly* 66(2): 267–279. DOI: 10.1177/1065912912442243.

Beck, G. (2013) Agenda 21 Keyword List. Available at: www.glennbeck.com/agenda21/agenda-21-word-list (accessed December 1, 2018).

Benjamin [pseudonym]. (2018) personal communication [phone call].

Bloomfield, E.F. (forthcoming) Ecocultural Identity in the Creation Care Movement: Analyzing Contemporary Performance of Religious Environmentalism. In: Milstein, T and Castro-Sotomayor, J (eds.) *The Routledge Handbook of Ecocultural Identity*. New York; London: Routledge.

Bloomfield, E.F. and Sangalang, A. (2014) Juxtaposition as Visual Argument: Health Rhetoric in Super Size Me and Fat Head. *Argumentation and Advocacy* 50(3): 141–156.

Bloomfield, E.F. and Tscholl, G. (2018) Analyzing Warrants and Worldviews in the Rhetoric of Donald Trump and Hillary Clinton: Burke and Argumentation in the 2016 Presidential Election. *Kenneth Burke Journal* 13(2): n.p. Available at: http://kbjournal.org/analyzing_warrants_bloomfield_tscholl (accessed December 1, 2018).

Brock, B.L. "A Definition of Four Political Positions and a Description of Their Rhetorical Characteristics," Diss. Northwestern 1965, pp. 335–354.

Brockriede, W. (1972) Arguers as Lovers. *Philosophy & Rhetoric* 5(1): 1–11.

Burke, K. (1969) *A Rhetoric of Motives*. Berkeley, CA: University of California Press.

Burke, K. (1984) *Permanence and Change: An Anatomy of Purpose*. Berkeley, CA: University of California Press.

Burton, T.I. (2014) Pope Francis's Radical Environmentalism. *The Atlantic*, July 11. Available at: www.theatlantic.com/international/archive/2014/07/pope-franciss-radical-rethinking-of-environmentalism/374300/ (accessed January 9, 2015).

Ceccarelli, L. (2011) Manufactured Scientific Controversy: Science, Rhetoric, and Public Debate. *Rhetoric Public Affairs* 14(2): 195–228.

Chamberlain, C. (1984) From Haunts to Character in the Meaning of Ethos and its Relation to Ethics. *Helios* 11(2): 97–108.

Charles [pseudonym]. (2018) personal communication [digital exchange].

Dillow, M.R. and Weber, K. (2016) An Experimental Investigation of Social Identification on College Student Organ Donor Decisions. *Communication Research Reports* 33(3): 239–246. DOI: 10.1080/08824096.2016.1186630.

Dunlap, R.E. (2013) Climate Change Skepticism and Denial: An Introduction. *American Behavioral Scientist* 57(6): 691–698. DOI: 10.1177/0002764213477097.

Elliott, K.C. (2014) Anthropocentric Indirect Arguments for Environmental Protection. *Ethics, Policy & Environment* 17(3): 243–260. DOI: 10.1080/21550085.2014.955311.

English Standard Version. (n.d.) Available at: www.biblegateway.com/

Fisher, W.R. (1984) Narration as a Human Communication Paradigm: The Case of Public Moral Argument. *Communications Monographs* 51(1): 1–22.

Fisher, W.R. (1985) The Narrative Paradigm: An Elaboration. *Communication Monographs* 52(4): 347–367. DOI: 10.1080/03637758509376117.

Goodwin, J. and Dahlstrom, M.F. (2014) Communication Strategies for Earning Trust in Climate Change Debates. *Wiley Interdisciplinary Reviews: Climate Change* 5(1): 151–160. DOI: 10.1002/wcc.262.

Hart, P.S. and Nisbet, E.C. (2012) Boomerang Effects in Science Communication: How Motivated Reasoning and Identity Cues Amplify Opinion Polarization about Climate Mitigation Policies. *Communication Research* 39(6): 701–723.

Johannesen, R.L. (1974) Attitude of Speaker toward Audience: A Significant Concept for Contemporary Rhetorical Theory and Criticism. *Communication Studies* 25(2): 95–104.

Kirk, K. (2018a) Climate Change Science Comeback Strategies. Yale Climate Connections, July 26. Available at: www.yaleclimateconnections.org/2018/07/climate-change-science-comeback-strategies-part-one/ (accessed December 19, 2018).

Kirk, K. (2018b) Focus on Those with an Open Mind. Yale Climate Connections, November 19. Available at: www.yaleclimateconnections.org/2018/11/focus-on-those-with-an-open-mind/ (accessed December 19, 2018).

Klumpp, J.F. and Hollihan, T.A. (1979) Debunking the Resignation of Earl Butz: Sacrificing an Official Racist. *Quarterly Journal of Speech* 65(1): 1–11.

Lakoff, G. (2010) Why it Matters How We Frame the Environment. *Environmental Communication* 4(1): 70–81. DOI: 10.1080/17524030903529749.

Long, R. (1983) The Role of Audience in Chaim Perelman's New Rhetoric. *Journal of Advanced Composition* 4: 107–117.

McFadden, B.R. (2016) Examining the Gap between Science and Public Opinion about Genetically Modified Food and Global Warming. *PLOS ONE* 11(11): e0166140. DOI: 10.1371/journal.pone.0166140.

Olivia [pseudonym]. (2018) personal communication [digital exchange].

Perelman, C. (1955) How Do We Apply Reason to Values? *The Journal of Philosophy* 52(26): 797–802.

Perelman, C. and Olbrechts-Tyteca, L. (1969) *The New Rhetoric: A Treatise on Argumentation.* Notre Dame, IN: University of Notre Dame Press.

Ratcliffe, K. (1999) Rhetorical Listening: A Trope for Interpretive Invention and a "Code of Cross-Cultural Conduct." *College Composition and Communication* 51(2): 195–224. DOI: 10.2307/359039.

Rosenthal, P.I. (1966) The Concept of Ethos and the Structure of Persuasion. *Speech Monographs* 33(2): 114–126. DOI: 10.1080/03637756609375487.

Ross, D.G. (2013) Common Topics and Commonplaces of Environmental Rhetoric. *Written Communication* 30(1): 91–131. DOI: 10.1177/0741088312465376.

Sangalang, A. and Bloomfield, E.F. (2018) Mother Goose and Mother Nature: Designing Stories to Communicate Information about Climate Change. *Communication Studies* 69(5): 583–604.

Segal, J.Z. (1991) The Structure of Advocacy: A Study of Environmental Rhetoric. *Canadian Journal of Communication* 16(3): n.p. Available at: http://cjc-online.ca/index.php/journal/article/view/626 (accessed February 12, 2018).

Shuman, E.K. (2010) Global Climate Change and Infectious Diseases. *New England Journal of Medicine* 362(12): 1061–1063. DOI: 10.1056/NEJMp0912931.

Tillery, D. (2018) *Commonplaces of Scientific Evidence in Environmental Discourses.* First edition. New York; London: Routledge.

Tormala, Z.L., Briñol, P., and Petty, R.E. (2006) When Credibility Attacks: The Reverse Impact of Source Credibility on Persuasion. *Journal of Experimental Social Psychology* 42(5): 684–691. DOI: 10.1016/j.jesp. 2005.10.005.

US Global Change Research Program (n.d.) Impacts on Society. Available at: www. globalchange.gov/climate-change/impacts-society (accessed January 2, 2019).

Wang, S., Leviston, Z., Hurlstone, M., Lawrence, C., and Walker, I. (2018) Emotions Predict Policy Support: Why It Matters How People Feel about Climate Change. *Global Environmental Change* 50: 25–40. DOI: 10.1016/j.gloenvcha.2018.03.002.

Wilchelns, H.A. (1972) The Literary Criticism of Oratory. In: Scott, R.L. and Brock, B.L. (eds.) *Methods of Rhetorical Criticism*, pp. 27–60. New York: Harper and Row.

3 Defining the bargainers

Revolution, modification, and negotiating with environmentalism

Although misaligned with the teachings of the Catholic Church, Galileo Galilei published *Dialogue Concerning the Two Chief World Systems* in 1632. His main argument was that the Earth was not the center of the universe, but that the correct spatial model was heliocentrism, or that the Earth rotated around the Sun. In the 1600s, Galileo's research was viewed as heresy because it contradicted prominent interpretations of the Bible. After many years, additional studies, and more experimentation, the scientific community and the Church accepted Galileo's interpretations of how celestial bodies move. I begin this chapter with a brief discussion of Galileo's excommunication because the process of negotiating between Christianity and new scientific information highlights the concepts of revolution and scientific discovery that underpin the bargainers' worldview. Unlike the separators, bargainers happily embrace climate science as a beneficial tool for understanding the environment. But, they place restrictions on what environmental science "counts" for decision-making, how it is used, and what can be interpreted from it from their Biblical perspective. This process of negotiation, where people adopt some aspects of climate science and modify others to fit their religious perspective, inspired the "bargainer" label.

Bargainers see themselves as skeptical, empirical, and scientific minds, rightfully supporting their interpretations of environmental data against mainstream and orthodox interpretations. Their identity as scientists, researchers, and informed individuals and their evaluation of climate science are core components of the bargainers' worldview. When bargainers make arguments and engage in conversation, they view the process as an exchange of equally valid and legitimate scientific expertise and experience. Bargainers strongly believe in their interpretation of climate science because they reached it through Biblical guidance. For bargainers, the Bible provides the lens through which science must be performed and through which the world must be understood. Without that originating lens, bargainers argue that other scientists cannot and do not see the full picture of science and its proper role in society.

It is important to discuss the bargainers after the separators to highlight the differences in the groups' discourses. While separators and bargainers are both religious climate skeptics and both reach conclusions contrary to mainstream climate science based on the Bible, they perform their skepticism and doubt through disparate worldviews and rhetorical features. These variations are of paramount importance when addressing and engaging climate skeptics. A self-proclaimed religious climate skeptic may fall into either category (or may do some strategic borrowing from both) and thus should be met with different approaches. While previous scholars (e.g., Matthews 2015) have distinguished skeptics based on the strength of their denial and the arguments they make, I situate bargainers and separators as having fundamentally different worldviews and guiding metaphors that alter how they respond to and engage information about climate change.

Before elaborating on potential strategies to engage bargainers in the following chapter, I first trace the bargaining and modification practices of exemplar groups exhibiting them. These rhetorical markers emerge from the frame of revolution and appear in the arguments that bargainers make against mainstream climate interpretations. The primary, definitive features of the bargainers' discourse is their guiding metaphor of revolution, which leads to rhetorical practices of constructing a Biblical filter, appealing to scientific standards, and cherry-picking data and experts. Bargainers' discourses are incredibly important to analyze because they appear, on first blush, to be reasonable, rational, and aligned with scientific norms. Bargainers may not be as loud, polarizing, or obviously opposed to climate science as separators, so they may be invited to participate in media interviews as reasonable counterparts to mainstream scientists (Dixon and Clarke 2013). Inviting bargainers to have a seat at the table might give the perception that the debate over climate science is still ongoing. Despite the risks in engaging separators and bargainers, having conversations might also open up opportunities for finding common ground and countering claims that alternative voices are being summarily and unreasonably shut out of climate change discourse. Navigating this tension between invitation and validation is a difficult communicative practice, but one that is integral to recognizing the presence and proliferation of bargaining strategies in mainstream discourse.

Based on the discourse of the Acton Institute (AI) and online bargainers, I outline typical bargainer strategies as characteristic of climate skeptics but that are distinct from separators. I invoke McGee's (1990) notions of "fragments" to collect samples of bargainers' discourse across groups and individuals. I focus on the discourse from the AI as the exemplar bargainer and supplement my analysis with interviews with individuals to contextualize the official discourse of these groups. Conversations were mostly held digitally or over the phone and were recruited through online requests.

The Acton Institute

The AI (n.d.-b) was founded in 1990 as the Acton Institute for the Study of Religion and Liberty, named after Lord John Acton. Their current president is co-founder Reverend Robert A. Sirico who wrote *Defending the Free Market: The Moral Case for a Free Economy*. Their mission is "to promote a free and virtuous society characterized by individual liberty and sustained by religious principles" (AI n.d.-c: para. 1). The AI primarily focuses on economic research and how free market policies, influenced by faith, are the proper foundations of a functioning democracy. The AI addresses diverse topics such as political campaigns, world leaders, taxes, culture, and the environment. Instead of activism, the AI doubts anthropogenic global warming and argues for a laissez-faire economic attitude to protecting the environment. In addition to publishing articles online, the AI also publishes books, a quarterly journal called *Religion & Liberty*, and a blog called the "Acton Powerblog." In addition, the AI funds student scholarships and hosts events, such as Acton University, which is a four-day lecture series that integrates religious teachings with economics (AI n.d.-a).

The AI shares some board members with the Cornwall Alliance (CA), our exemplar separator group, but operates as a separate, research institution committed to producing discourse to halt current climate change policies. The borrowing of staff between the separators and bargainers can easily be explained by their shared skepticism of mainstream climate change conclusions. Similar to the CA, the AI argues that current environmental policies will negatively affect the economy and the poor, but their rhetorical features and argumentative strategies are different. Instead of villainizing or demonizing environmentalists, the AI argues that some politicians are unduly influenced by environmentalists and should instead be more concerned about sound economics. While there may be topic overlap between separators and bargainers in terms of what arguments they make (i.e., claims of climate science being a religion or claims of climate science being a scam), this overlap is not mirrored in their guiding frames nor their rhetorical features. Because my typology is based on worldviews and underlying perspectives, it allows for categorization even when people may make use of a variety of climate skeptical arguments. The next section outlines the guiding terms for the AI and bargainers in how they frame climate science as lacking sufficient research, ignoring Biblical and economic considerations, and undergoing a revolution sparked by climate skeptics.

Guiding terms

The bargainers' primary goal is to correct what they perceive to be fraudulent conclusions from climate science data by offering competing interpretations. Bargaining groups are not fighting a war; instead, they are revolutionaries opening the eyes of the scientific orthodoxy to more accurate information about and interpretations of reality. Bargainers try to shift dominant environmental science conclusions toward faith-based science, where the Bible serves as the basis for scientific claims about climate change. In alignment with Thomas Kuhn's "scientific revolution," bargainers seek acceptance of their minority scientific opinion that they believe will eventually become the mainstream scientific standard, thereby reversing previous conceptions of how humans impact the environment. Those I label bargainers align with Ceccarelli's (2011: 198) argument that skeptics identify themselves "as heroes in an unfolding scientific revolution" whose opinions are viewed as "heresy" by the scientific mainstream.

In the narrative bargainers tell and the frame through which they act, they are bolstered by their persona as revolutionaries and whistle-blowers in the scientific community. Because they view themselves as part of a legitimate scientific, informed community, they often do not engage in the more hostile, vitriolic rhetoric of the separators. Instead, the bargainers' revolution works within modern climate science to make room for religion as a legitimate, and even more valid, way of knowing about the Earth and the environment.

Kuhn (1996: 2) argued that notions of science developing through a "piecemeal" process, by which new information and evidence expands upon existing knowledge, are not always accurate. While scientific knowledge does build on previous knowledge, this process is neither exclusively linear nor progressive. In some instances, new information replaces or displaces previous paradigms and conceptions of interactions, causing rifts in the scientific community between old and new conceptualizations that need to be repaired and reexamined before adopting the new paradigm. Kuhn (1996) argued that science's simultaneous arbitrariness and rigidity create the conditions for shifts in established information and methodologies. In other words, what becomes known to be science is the current consensus on the best answer and explanation at the time, which is designed to shift and change as new information is discovered.

Within climate science, the bargainers position themselves as revolutionaries who are attempting to institute "a new basis for the practice of science" in order to "shift" current scientific standards (Kuhn 1996: 6). They view current mainstream climate scientists as the old paradigm, wedded to an outdated perspective in need of change. Those who support the old paradigm are viewed as unwilling to view the valid scientific evidence the bargainers put forth as a counter to the anthropogenic climate change paradigm. By substituting their own evidence and positioning themselves as contributors to scientific knowledge, the bargainers seek to highlight "anomalies that subvert the existing tradition of scientific practice" so that science "can no longer evade" the truth of their brand of climate science (Kuhn 1996: 6). The bargainers, therefore, do not need rhetorically to start a war; they only need to create "competition between segments of the scientific community" in an attempt to promote the "adoption of another" scientific standard, or at least, show that there remains vocal opposition to climate change from within the scientific community (Kuhn 1996: 8).

The AI and other bargainers forgo the metaphor of war in favor of revolution, and not necessarily a violent one. Their language is sometimes aggressive or belligerent toward identified opponents, but it is not as polarized or dramatic as the separators. Instead, the bargainers characterize their enemies as misguided and in need of replacement. Unlike the separators, who are waging a moral crusade against their enemies, the bargainers see redeeming qualities in climate science, its methodologies, and conclusions, when performed correctly. The bargainers seek to adopt some of climate science's qualities that fit their worldview to reclaim space for religious interpretations. The bargainers borrow from science's esteem and position of power to reinforce the legitimacy of religious interpretations of that information. In positioning themselves as a legitimate part of the scientific community, bargainers wish to change the dominant ideas, not under-mine the entire institution. Although the term bargainers might imply the reach-ing of a middle point or compromise between their Christian faith and climate science, bargainers primarily negotiate and adapt the tenets of science to fit their religious worldview (while harmonizers will do the opposite). Bargainers thus stretch, adopt, and borrow some features of climate science and abandon others to maintain their religious foundations.

When mainstream scientific conclusions and one's interpretation of religion appear to oppose one another, religious adherents may struggle to consider both as authentic and valid sources of truth. To incorporate and make sense of that information, they may practice a weighing and blending of the two, which Burke (1984: 23) called "casuistic stretching." When undergoing casuistic stretching, old ideas may be modified to fit new information, or new information may be interpreted to fit previous held worldviews. Because religion features so promin-ently in people's identities and sense of self, those beliefs may be particularly hard to stretch, especially if one feels that inherent values and moral convictions are at stake.

Because of the nuance in the bargainers' narrative, they do not fit easily into any one frame. In Chapter 1, I discussed the melodramatic frame that emerges in separators' discourse. Bargainers, although they are also climate skeptics, do not turn to melodrama, but to a blurring between the comic and tragic frames referred to as a tragicomic frame. Burke (1984) called comedy and tragedy "frames of acceptance" where people work within the existing system. Tragic frames promote the perspective that pollution in the world must be sacrificed to restore the order. Comic frames promote the perspective that pollution is largely the result of error and misfortune, rather than human malice. For the bargainers, they tend to adopt a tragicomic perspective defined by Smith and Hollihan (2014: 585–586) as a "Burkean Serenity Prayer," where we "accept the things we cannot change," but also identify ways that things can and should be changed. Bargainers adopt elements of the comic frame in that they do not view contemporary environmentalists as villains; they do not engage the separators' holy war. Instead of seeking to vanquish opponents, bargainers wish to correct and set right those that have misinterpreted scientific evidence. The comic frame is characterized by its insistence that in conflict, people are "mistaken" instead of malicious (Burke 1984: 41). Desilet and Appel (2011: 346) argued that the logical extension of the comic frame's tendency to see enemies as "mistaken" is that comic frames "are thereby prevented from seeing one side as exceptionally and accountably wrong." Schwarze (2006: 242) agreed and argued that the comic frame "forsakes the divisiveness of the tragic frame." The bargainers adopt this element of the comic frame and enact an argumentative approach to mainstream climate scientists and environmentalists.

Although bargainers evoke some elements of the comic frames, they also contain tragic features, such as their condemnation of some features of environ-mentalism. But, bargainers' rhetoric diverges from a pure comic frame in that they clearly do invoke divisiveness through the process of differentiating between themselves and environmentalists. Quoting Burke, Desilet and Appel (2011: 348) emphasized that frames can evoke factional conflict, which "attributes the evil, not to all men [*sic*], but to some." In identifying some features of climate science and some climate scientists as oppositional to the bargainers' message, the bargainers' frame "shifts away from the apparent culpable egalitarianism of the comic frame toward the moral polarity of factional weighting" more common of the tragic frame (Desilet and Appel 2011: 347).

The bargainers' narrative thus appears to collapse frame categories, where not only are there specific, identified enemies that are responsible for the current paradigm and must be overthrown (tragic), but also there is an attempt to revolutionize and change the field of science from those who are mistaken (comic). The bargainers' guiding frame is thus most accurately classified as a factional tragicomic frame. The bargainers accept that climate science is an institution of power and authority that provides valuable information about the world around us. In accepting this viewpoint, however, bargainers also acknowledge that they have the power to change and modify what are considered accurate interpretations and conclusions. That is, they accept the validity of science as a way of knowing but attempt to challenge its exclusion of religious grounds and evidence from climate change decision-making. Desilet and Appel (2011: 356) proposed that sometimes "the comic framer" may have "to adopt rhetorically tragic structurings of conflict" to communicate their message. The bargainers appear to be flipping this dynamic, using comedic language of mistakes, correction, and education to cover a tragic interpretation of how the order should be restored – through replacing current climate science with the bargainers' version of climate science that is filtered through a Biblical worldview.

The differences between frames may appear trivial. But, the adoption of a frame changes the entire narrative and thus appropriate actions and discourse in response to it. Scholars have noted that a shift from tragic to comic or comic to tragic transforms nearly all aspects of a rhetorical moment, including audience reception and message success (e.g., Carlson 1986; Christiansen and Hanson 1996). In this vein, the shift from the separators' melodrama to the bargainers' tragicomedy indicates a significant change in their guiding framework and subsequently their treatment of environmentalists and climate science. The bargainers start with the Bible and propose that their methods of inquiry and Biblical evidence provide additional knowledge to existing climate science research. The ongoing discussion (much like the separators' ongoing war) legitimizes the need for the bargainers to exist and continuously address climate change. Bargainers reopen the debate to infuse a new purpose: legitimizing their own explanatory narratives, religious values, and priorities in environmental decision-making, which the reigning environmental paradigm has incorrectly abandoned. In what follows, I outline three prominent rhetorical features of the bargainers resulting from their revolutionary metaphor and their tragicomic frame: (1) constructing a Biblical filter, (2) appealing to scientific standards, and (3) cherry-picking evidence and experts.

Constructing a Biblical filter

Part of the bargainers' core identity is that they interpret climate information through a Biblical worldview that is often accompanied by a mediating topic, such as economics. While climate science often relies on the consensus of climate scientists as evidence of climate change's inevitability, bargainers discount such arguments because they distrust the credibility and impartiality of climate scientists who depart from their interpretations. Part of this distrust comes from mainstream climate scientists removing the Bible from consideration. For many bargainers, their faith does not simply indicate a reinterpretation of climate science, but their faith also structures their political and economic worldviews, which themselves are intertwined with issues of environmental advocacy and proper environmental policy. Similar to the separators' strategy of controlling definitions, the bargainers' Biblical filter aims to delineate between what are appropriate interpretations of data and what should be discarded. Furthermore, they insert political and economic factors as equally important considerations and competing inventional resources to climate science. Unlike controlling definitions, bargainers are less concerned with proper terminology than proper interpretations and what viewpoints should have a seat at the table. To view their arguments through a Toulmin model, bargainers may reason from the same evidence as a climate scientist or an environmentalist but will use them to support a different claim. The grounds given are the same, but the authorizing warrant will not enable a claim that contradicts the Bible or their political and economic loyalties. Given the same grounds, the models may appear as below:

Mainstream climate science model

Grounds: Global temperatures are increasing as can be seen in Intergovernmental Panel on Climate Change (IPCC) data.
Warrant: The IPCC is a reputable, credible source of information that should guide policy decisions.
Claim: We should act now to make policy changes to prevent the consequences implied by these temperature increases.

Bargainers' climate science model

Grounds: Global temperatures are increasing as can be seen in IPCC data.
Warrant: Science is only part of the equation, we must also consider the Biblical and economic implications of environmental politics.
Claim: We need to do more research and consider multiple perspectives before acting on existing climate data.

In the second Toulmin model, we can see how bargainers can draw from the same evidence as climate scientists, but interpret that data as supporting a call for more research instead of proof of validated, conclusive information. The exemplar bargainer, the AI, frequently uses calls for more research to promote hesitancy and delay in environmental decision-making, a documented strategy of climate skeptics (Ceccarelli 2011).

The AI defines scientific, economic, and environmental understanding as resting on Biblical principles. In an AI public policy report, the AI argued that environmental policies should be based on "scientific understanding built on a biblical worldview" (Beisner *et al.* n.d.: para. 43). The Bible serves as the foundation from where science emerges; without that foundation, the AI views science as unrooted and inaccurate. Beers *et al.* (n.d.: para. 5) echoed the theme of construction and building when arguing that "an environmental ethic ... rests firmly upon the foundation of both sound reasoning and divine revelation." These building metaphors are echoed in the Bible, which frequently makes reference to the importance of foundations in creating the Church. For example, 1 Corinthians 3:11 reads, "For no man can lay a foundation other than the one which is laid, which is Jesus Christ." The AI views Christianity as a starting point from which they build their environmental arguments. A strong foundation is important because a structure built atop it can withstand rain, floods, winds, and other challenges and not fall (Matthew 7:24–7:25).

Ballor (2006: para. 8), a contributor to the AI, described economic arguments as working hand-in-hand with faith to justify decision-making, noting: "economics helps us rightly order our stewardship." Because the AI connects free market ideologies with Christianity, a reliance on economics over environmentalism is analogous to weighing Christian values over perceived non-Christian values. Ballor (2006: para. 7) argued that the terms economics and stewardship come from a "shared biblical origin" through the term *oikos*, meaning house or order. Ballor (2006: para. 4) argued that the term economics stems from *oikos*, which also refers to management of the household, which can subsequently be translated as "steward" or "manager." This common origin should encourage people to "see them as united" instead of separate, as Ballor (2006: para. 8) accuses some religious environmental groups of doing.

The use of origin is important here, because it reveals how the AI sees all things as rooted in the Bible. For the AI, their faith requires equal consideration of both economics and the environment, where the former dictates the proper ordering of the latter. In another use of the term origin, Beers *et al.* (n.d.: para. 1) argued that the AI's work "derives its authenticity from its origin, which is Christ himself." The AI views the Bible as an "indispensable point of departure" from which it gains a "deeper understanding of the created order ... [and] humanity's value and place in that created order" (Beers *et al.* n.d.: para. 2). The Bible is a foundation, origin, and argumentative starting point that guides subsequent thought and reasoning. Based on their interpretation of the Bible, the AI creates a formal hierarchy where social and economic issues outweigh environmental ones. For example, the AI accuses environmentalists of abandoning "the most pressing areas" of Christian concern such as helping the poor, controlling disease, and improving trade (Ballor 2006: para. 12). Environmentalists thus disrupt the proper order. In order to restore order, the AI has to shift the emphasis of environmentalism and mainstream climate science and replace it with Biblically-oriented climate science and economic decision-making. The AI argues that contemporary environmental policy should "take into account everything that the sciences … are able to tell us about our world" (Beers *et al.* n.d.: para. 56). But, the information that science provides is only valid if it "integrate[s]" with a Biblical interpretation and can be interpreted to match "the normative principles of the moral order" (Beers *et al.* n.d.: para. 56). Both of these statements work to emphasize stretching climate science to fit the AI's interpretation and work to subsume the environment under other concerns.

Similar to the CA, the AI emphasizes caring for the poor and helping humans over the environment. While science and economics are viewed as "tools that allow us to respond to the multifaceted problems we face," AI authors note that "science alone is insufficient for resolving these matters," and must be aided by faith (Beers *et al.* n.d.: paras. 62, 63). The AI does not view mainstream scientists who side with environmentalists as evil, but consider environmentalists and climate scientists to have been distracted by the environmental message and as a result may endanger humanity's future. The AI proposes an economic, free market response to perceived environmental problems. For example, AI contributor Phillips (2010: para. 5) argued that "the public-policy response to global warming proposed by some evangelicals makes actually helping the global poor more difficult." Christians that support restrictive policies are thus working against the AI's initiatives and other commitments of the Christian "social ministry" (Phillips 2010: para. 3). The AI engages in "proper environmental stewardship," defined as putting "human needs above non-human needs when the two are in conflict" (Beisner *et al.* n.d.: para. 7). This statement clearly articulates a strict hierarchy of human economic autonomy over environmental stewardship. The AI references the Bible, "God gave men and women superiority and priority over all other earthly creatures" (Beisner *et al.* n.d.: para. 7) to support claims to human's superiority and dominion over nature.

In my ongoing conversation with Abigail she noted that climate change was often used as an overarching term to capture various connotations. She was concerned that "climate change" as an imposing, all-encompassing topic was distracting people from reality. Similar to the discourse about hierarchy used by the AI, Abigail was concerned about how to best care for the environment without imperiling anyone's financial wellbeing. In other words, money and autonomy were ranked as more valuable than environmental advocacy. Calling regulations such as the Paris Accord a "financial strain" on individuals and businesses, Abigail wanted "responsible stewardship" to be interpreted through a Biblical lens of helping society by "dramatically reduc[ing] pollution." Curbing pollution was as far as she initially went in acknowledging the environmental issues currently facing the world. Abigail summarily dismissed anthropogenic global warming as a "scam" and something that is exaggerated in order to claim attention and money for personal gain, rather than tackling issues she could see and observe in her own community. These arguments show how intertwined Abigial's economic and religious perspectives are and how the two ideologies work together to determine proper environmental actions. Another bargainer I spoke with, Joseph, argued that he is not really a skeptic because he believes in climate change. He noted, "The climate is changing. It's been changing for millions of years and will continue to do so." By breaking apart "climate change" into the general idea of a changing climate, and not the accurate definition of severe anthropogenic warming, Joseph gives the impression that he believes in climate science, while at the same time heavily modifying the meaning of "climate change" to fit his narrative that nothing needs to be done.

In addition to modifying the term climate change, bargainers tend to modify the stakes of climate change conversations. For example, the AI conceptualizes policies that aim to help the environment and the poor as a zero-sum game instead of mutually beneficial: "All of this [environmental advocacy] involves material costs, resources devoted to this advocacy that might also be used else-where" (Ballor 2015: para. 5). For the AI, to prioritize the environment, then, is to choose against the poor and other humans in need. This discourse cuts against the prevailing environmental discourse of climate justice, which argues that it is the poor who will suffer most from disrupted global systems and climatic changes from increased warming (de Onís 2012). Climate justice, in the eyes of the AI, ignores issues of justice, morality, and Christianity's "broad mandate that includes many issues other than climate change, including abortion" (Ballor 2015: para. 5). For the AI, causes such as environmentalism, the pro-life move-ment, and charity to the poor compete with one another for attention and money. Environmentalists' lack of faith and understanding of economic theory leads them to misinterpret their "good intentions" as "morally good ends," when real Christians, such as the AI, would not consider them so (Jensen 2015: paras. 6, 3). Labeling climate scientists as having good intentions enacts comic framing where environmentalists are not evil or malicious, but misinformed. Yet, the AI does condemn these interpretations as separate, distinct, and oppositional to their own, invoking features of a tragic frame.

For example, the AI characterizes current conversations about climate change as falling short of the AI's standard of "sound environmental stewardship" that is "wiser and more biblical than that of mainstream environmentalism" (Beisner *et al.* n.d.: para. 4). Placing mainstream environmentalism in opposition to the AI's work implies that mainstream science is unsound and inaccurate in its con-clusions about the environment. Correct environmental actions must be "bibli-cally sound" and thus mainstream climate science is unqualified to deliberate on them (Beisner *et al.* n.d.: para. 11). An action is Biblically sound when it matches the teachings of scripture and resonates with predetermined Christian values. By using vocabulary such as "sound" and "unsound," the AI evokes ideas of unsteady and fragmented foundations upon which current climate science stands. Any actions must meet the standards of sound religion and science, both as interpreted by the AI. The AI undermines the current under-standing of climate science by questioning whether mainstream science is sound science at all.

The AI's arguments are built on the foundation of economic freedom, God's decision-making abilities, and the inerrancy of the Bible. The AI lauds economics, in particular, as being an integral aspect of proper Biblical behavior: "Economic activity [is] an extension of God's own wisdom for how man is to relate to his physical surroundings" (Beers *et al.* n.d.: para. 68). A Biblical approach to the environment thus cannot be separated from "the freedom and responsiveness of markets" (Beers *et al.* n.d.: para. 68). For the AI, the most prudent environmental actions appear to be no action at all, as the autonomy of the market will instead self-correct environmental issues. The AI argue that there is no need to rely on environmental science, because the Bible contains theological guidance for "all environmental questions" (Beers *et al.* n.d.: para. 2). To stray from this resource is to succumb to the "insufficiency of human reason" as a subpar substitute and reorder the proper hierarchy of humans and economics over the environment (Beers *et al.* n.d.: para. 3).

Appealing to scientific standards

Perhaps most unnerving about the bargainers is their persistent appeals to scientific standards to justify patently unscientific conclusions. Unlike the overt hostility that often comes from separators' discourse, bargainers appear to hide in plain sight, in that many of their doubts and arguments may appear to be legitimate and valid concerns. My characterizations of bargainers share many similarities with skeptics who aim to manufacture controversy (Ceccarelli 2011) over the status of climate change knowledge. Bargainers, in painting themselves as part of the scientific community, seem to be engaging in productive conversations and deliberations. But, considering the established consensus on climate change, their concerns seek to halt discussion about appropriate actions and policies by re-engaging arguments for proof and clarification (Ceccarelli 2011). In this section, I highlight how bargainers stretch and borrow scientific standards by appealing to open-mindedness, weighing all sides of the issue, and calling for more research.

In her discussion of manufactured scientific controversies, Ceccarelli (2011: 198) argues that skeptics appeal to "open-mindedness, freedom of inquiry, and fairness" to create "discursive traps to constrain the responses of mainstream scientists and their allies." Similar to the skeptics Ceccarelli (2011) describes, the AI calls for a dispassionate, apolitical discussion about the environment, where faith, economics, and only some climate science are invited to the table. Beers *et al.* (n.d.: para. 63, emphasis added) argued, "We need the very best and *dispassionate* environmental science" to guide current actions, implying that current environmental science is biased. The AI accuses environmentalists and leftists of having misconstrued scientific inquiry into climate change and having "undermined the quality of debate over both science and public policy" as a result (Beisner *et al.* n.d.: para. 75). The AI argues, "we desperately need an *authentic* democratic deliberation on the environment" before any "real environmental decisions" can be made (Beers *et al.* n.d.: para. 67, emphasis added).

Such phrasing encourages hesitation and delay on environmental decision-making shielded by the appearance of concern for accuracy, impartiality, and deliberation. Beers *et al.* (n.d.) implied that authentic democratic deliberation has not yet occurred, but the AI is the group that can provide it. The AI described its interpretations as "sound science rooted in a value structure that emphasizes honesty and openness to debate and evidence," again implying that these qualities are not features of mainstream climate science (Beisner *et al.* n.d.: para. 75). The AI accuses environmentalists of having "derided the motives of scientists and others who questioned that conclusion [of scientific consensus]" (Beisner *et al.* n.d.: para. 75). Indeed, the AI advocates that "the science of climate change is not decided," in part because of the closed-mindedness of mainstream scientists (Snow 2015: para. 14).

The AI cautions against hasty action because of a lack of information and consensus on the proper response to existing information. An AI brief concluded, "policy makers should be *very slow* to base their decisions on model predictions" (Beisner *et al.* n.d.: para. 62, emphasis added). In addition, Beers *et al.* (n.d.: para. 67) argued, "we must proceed with great caution and prudence" in making "real environmental decisions." The AI encourages further research and modeling before drastic action is taken to protect the environment. Due to "the highly uncertain nature of both the theory and the evidence of global warming," the AI proposes "to delay action … until the matter is much better understood" (Beisner *et al.* n.d.: para. 74). The AI perpetuates a focus on whether or not anthropogenic climate change is happening by positioning itself as a dissenting scientific institution with opposing evidence and interpretations of that evidence.

Ceccarelli (2011: 198) argued that those who manufacture scientific controversy paint mainstream science as "dogmatically unscientific" as a strategy to borrow their legitimacy while undermining their intended meanings. For example, the AI characterized climate science as more closely related to religion than to the AI's standards for science. Beers *et al.* (n.d.: para. 80) argued that some people "mak[e] idols of nature or creatures that, in doing so, exalts them above our primary duties toward God." In this way, environmentalists are members of a faith "who worship the created, but not the creator" (Snow 2015: para. 5). This perspective is "wrong-headed and dangerous," because it goes against God's intentions and leaves nature "untamed" (Beers *et al.* n.d.: para. 45). Furthermore, this perspective devalues human life as "a drain on resources" (Beers *et al.* n.d.: para. 50). The AI does value the environment, but not at the cost of human life or going against God's orders. These ideas were echoed by Abigail, who noted, "Global warming seems to be a secular religion for many, which is ironic because of the faith required from a group that seems to be more focused on logic." If the bargainers are more scientific than climate scientists, then climate science must be something other than science. This rhetorical strategy paints climate science as lacking in its own scientific standards and argues that science must return to those standards in order to realize the truth.

Abigail thought it was amusing that climate scientists were so alarmed about the environment and were pushing what she considered to be "disgustingly corrupt" policies under the shield of "science." She further surmised that "God is quite entertained at the frivolous wisdom of the humans trying to save the planet," because their science is moving them in the wrong direction. Abigail noted that she does not see "a conflict between religious belief" and "climate change," but believes that scientists are interpreting the data for climate change incorrectly. When I directly asked Abigail about the scientific consensus about the existence and severity of climate change, she replied, "I do not believe the science is settled at all, and the 97 percent agreement statistic is also false." To support this conclusion, Abigail linked to three online articles, two from Forbes contributors and a report published by Friends of Science about the manipulation present in the reporting of a 97 percent consensus (Friends of Science Society 2014). Unlike my conversations with separators, bargainers were far more likely to link to external sources and provide competing evidence and interpretations of that evidence. This also points to the fact that bargainers will invoke scientific authorities, but only those that provide alternative interpretations to mainstream climate science and align with their existing beliefs on climate change.

In a conversation with Victor, he argued that climate science was getting ahead of itself in proposing policy changes such as "carbon offsets." Instead of jumping to what he called "the third step" of policy changes, science must first "prove there's actually a problem" and second "find the root cause of the problem." Only "once all of that's figured out and actually true" can policy discussions take place. Victor argued that climate change is packaged in a way to engage everyone in its importance "without a lot of proof." This argument mimics stasis theory, which is an argument procedure that posits we must first agree on the existence of an issue before working to define, evaluate, and solve it. Through using this argument, Victor appeared logical, systematic, and thoughtful. He believed that there is not enough evidence in step one or two to justify making any changes before additional research is done. Furthermore, Victor seemed to doubt whether such recommendations were appropriate for science to make. In distinguishing what science should be able to do and not do, scientists are restricted to having credibility and establishing ethos only over certain matters of science unrelated to policy. Walsh (2013: 164) highlighted this issue in her chapter on the IPCC that described how IPCC reports gradually moved from making climate change "a scientific dilemma to a political one." In making that rhetorical shift, the IPCC was accused of having "transgressed the is/ought boundary by making value judgments" (Walsh 2013: 166). In other words, the IPCC report did not end with telling us the facts, the status, the "is" of climate change, but went further in telling us what we should and "ought" to do about it. In making this shift between scientific and political discourse, the IPCC's ethos was under scrutiny over being capable of making what this bargainer called a step three claim.

Bargainers strategically locate climate change discourse in early parts of stasis theory that focus on existence and definition to stall deliberation over appropriate actions. Bargainers validate their skeptical perspectives not only by winning the debate, but also by "delay[ing] policy changes" and thus keeping the debate, and their voice in it, alive (Ceccarelli 2011: 197). On its surface, this argument seems fully rational and reasonable; we must start at the beginning before we move through to more advanced issues. Fahnestock (1986: 345) argued that science communication naturally operates via stasis arguments, first by convincing its audience "that a situation exists before they ask what caused it or move to decisions" about it. But, scholars warn that keeping the debate in the realm of definitions, facts, and forensic rhetoric inevitably prevents true progress from being made (Bloomfield and Lake 2015; Ceccarelli 2011). For example, in a conversation with Victor, he discussed his nuanced opinion about his skeptical status and self-identified as a "lukewarmist" who acknowledged some of the evidence of CO_2 increases, but saw "little evidence" to support "climate activism" in response to this information. This agreement was shared across bargainers I spoke with, who would readily acknowledge that increased CO_2 and temperatures have occurred but did not make the interpretive leap that the evidence about these increases requires action to be taken.

In addition to emphasizing the value of and need for more research about climate change, another scientific standard that bargainers appeal to is that science is a malleable, changeable, and constantly evolving practice. Operating under the metaphor of revolution, the AI is confident that future science will confirm their conclusions. Beisner *et al.* (n.d.: para. 38) described the environmental assumption "that as people grow in numbers, wealth, and technology, the environment is always negatively affected … is falsified by hard empirical data." The AI argues that "highly speculative computer climate models drove the great fears of global warming" but actual measurements undercut estimates of carbon dioxide and warming (Beisner *et al.* n.d.: para. 60). The AI argues that environmentalism will eventually be replaced and overthrown by the AI's interpretations of climate data. AI contributor Ray Nothstine (2015: para. 6) argued that present-day Christians are not unlike how Christ himself was "on the wrong side of history" against the faith of the Roman Empire. With time and more research, the AI asserts that they will and replace the current environmental focus with conservative Christianity-based climate science.

One of the bargainers I spoke with shared the AI's confidence that mainstream climate interpretations would be displaced. Abigail seemed content to let conversations about climate change progress as they naturally had been because discussion and deliberation would reveal that the climate change hysteria was overblown. If society can keep pouring over the data and double-checking information and models, then the new norm for mainstream science would become skepticism toward humanity's impact on the environment. This bargainer did not have the language of Kuhn's scientific revolution but was referencing the same concept that science sometimes changes course and meanders in unexpected ways. Our exchange contained few aggressive or warring terms; Abigail was content that she had figured out the truth and that "real" science justified her position. There was no need for aggression or war because the revolution was imminent. Whether intentional or not, the lack of aggression in the bargainers' worldview also grants them the air of being dispassionate, and thus, scientific.

The AI goes even further and argues that not only will mainstream science eventually agree with the AI's conclusions, but that society will also use the economy and free market environmentalism "to provide ways to solve many other problems we currently face" (Beers *et al*. n.d.: para. 60). The AI appeals to the idea of Energy Darwinism (Channell *et al*. 2013, 2015), where further innovation will encourage adaptation and richer solutions to environmental problems without a need for regulation and government intervention. Beers *et al*. (n.d.: para. 60) further noted, "human societies require greater development and more innovation, not less," so investment in new technologies should be encouraged, thereby solving both environmental and economic concerns. An investment in these new technologies confirms the AI's beliefs that "democracy and markets ... are the best mechanisms for the responsible handling of the environment" (Beers *et al*. n.d.: para. 61). By encouraging reliance on markets, the AI shifts focus away from regulatory policies and onto the economy as environmental solutions. Future research will confirm the AI's conclusions, putting them on the forefront of scientific knowledge and innovation.

Appealing to scientific standards provides bargainers a seat at the table. Being members of the scientific community and sharing in science's values affords bargainers the presumption of rationality and good intentions. By appealing to scientific standards, bargainers turn the tables on climate scientists as discriminatory for excluding their voices and undermine climate scientists' commitment to open dialogue with those that disagree with them. Instead of directly rejecting climate science, as some separators do, bargainers seem more level-headed and open-minded in considering scientific data and requesting more information before making a decision. The bargainers' strategies appear to be largely successful at extending conversations and keeping dialogues about climate science's accuracy in the public eye to distract from the urgency of reaching solutions.

Cherry-picking evidence and experts

Although bargainers appear to appeal to scientific standards, they diverge from contemporary scientific practices in important ways. For example, bargainers often appear to be using valid scientific evidence, but instead are selectively representing and cherry-picking from the full picture of mainstream climate science. Ceccarelli (2011: 197) argued that scientific uncertainty is in part manufactured by taking scientific information "out of context," using "cherry-picked" data, and pointing out flaws in mainstream studies that have "inconvenient results" for the alternative narrative. Bargainers often selectively pick sources of authority or pieces of evidence that align with their beliefs about the environment but are still validated and credentialed parts of the scientific community. Bargainers seek out people and data that support their views among the many in climate science that do not. These voices are often representative of a minority that continues to challenge established climate change facts despite the overwhelming majority. The lack of evidence in the mainstream does not deter the bargainers, however. Instead, the dearth of support only bolsters their insistence that their position is the rightful one being quashed and unscientifically silenced by contemporary scientists and elites. Highlighting specific data and lauding minority voices plays into the bargainers' revolutionary worldview. The metaphor of revolution provides a lens through which all voices and all data are important and could potentially provide field-changing information. Kuhn (1996: 6) noted that sometimes extraordinary shifts happen after small results, because "anomalies that subvert the existing tradition of scientific practice" trigger reframings of entire perspectives and approaches to doing science. For bargainers, the fact that there are any dissenting voices is enough to provide pause and make space for those alternative viewpoints. Because bargainers' view of proper climate science is seen through a Biblical (and oftentimes economic) lens, they are particularly open to evidence and voices that emphasize or align with Christian and neoliberal values.

For example, Victor was convinced that climate scientists manipulated data and did not share the full picture of climate change. He argued that climate activists and journalists often portray the "worst case scenario" as the most likely scenario. In other words, Victor argued that climate activists and journalists may exaggerate the predicted models to increase attention and sensationalism. Although the vast majority of climate models predict disaster, Victor infers manipulation on the part of climate activists who unscientifically exaggerate and inflate urgency and severity to pull people toward their perspective. Victor argued that if you really look at the data, the more likely scenarios are not hysterical or apocalyptic.

Joseph shared this skepticism of academics and elites who are not open to hearing all perspectives and are inserting environmental bias into the conversation. He noted that dissent has been "suppressed" in academic communities, specifically on the topic of climate science and attributes an ulterior motive to such actions. Joseph was particularly wary of the influence of money, not in businesses, but by environmental interests, on information and data that is published. He noted that he was oftentimes skeptical of engaging in climate change discourse in general because he felt he could not "find unbiased information" that was not tainted in some way. Joseph also described his concern and confusion about "which PhD is the right one." Joseph described his difficulty in doing research and being faced with extremely smart people who all seem to be credentialed and know what they are talking about, so was uncertain how to distinguish between them. this conversation provides evidence that the presence of minority viewpoints, especially from mercenary scientists who hold the appearance of credibility and expertise on topics of climate change, can be confusing elements within climate change discourse and complicates general advice that the public should listen to and follow scientific authority.

From the bargainers' perspective, mainstream climate scientists unnecessarily and wrongfully silence dissident voices that go against the paradigm of severe anthropogenic climate change. To counter these silencing behaviors, bargainers refute published works and try to reinterpret their conclusions. For example, bargainers tend to refute claims that CO_2 is damaging to plants. McCright and

Dunlap (2003: 351) call this strategy the "social construction of non-problematicity," whereby "negative environmental conditions" such as rising rates of CO_2 are reframed as "non-problematic" and even beneficial. The AI wrote a report arguing that "some benefit is to be expected – indeed, has already occurred – because of enhanced atmospheric CO_2," such as plant growth and farming productivity (Beisner *et al.* n.d.: para. 66). This article cites scientific reports that plants have enhanced growth under conditions of high CO_2. In this sense, the AI is borrowing the voices of scientific authorities to refute mainstream climate science's argument that climate change and increased greenhouse gases are harmful and thus concerning. While the report cited is part of mainstream scientific data, the AI reaches markedly different conclusions about the level of CO_2 at which the benefits cease and plant life is adversely affected by the changed atmospheric make-up. In picking and choosing which data and which interpretations count as authentic and real, the AI constructs a strategically curated view of climate science that supports their Biblical and economic values. This feature of climate denial is a prominent one; previous scholars have identified skeptics' commitment to scientific standards (Bloomfield and Tillery 2019; Ceccarelli 2011) as a strategy to adopt the credibility and legitimacy of science without following all of its tenets or equally considering evidence for an opposing perspective.

The AI frequently references their "Global Warming Petition" as a resource to reinterpret the oft reported 97–98 percent consensus of scientists that support anthropogenic climate change (e.g., Anderegg *et al.* 2010; Bray 2010; Cook *et al.* 2013; Doran and Zimmerman 2009). The Global Warming Petition has been signed by "more than 17,000 basic and applied American scientists," which the AI argues constitutes a significant dissensus in climate science (Beisner *et al.* n.d.: para. 81). The AI strategically uses the raw number of signatories rather than discussing the percentage to increase the appearance of dissent. In other words, can 17,000 scientists be incorrect? The statistic of nearly 20,000 scientists sounds like a fair number of dissenters, but this number should be compared to and put in conversation with the number of scientists who do believe in anthropogenic climate change, making the question, can millions of scientists be incorrect? National Aeronautics and Space Administration (NASA) lists the current American scientific societies and institutions that collectively have made statements endorsing anthropogenic climate change, including the American Association for the Advancement of Science, the American Meteorological Society, the US National Academy of Sciences, and the IPCC (Shaftel *et al.* 2018).

In addition to reinterpreting data, bargainers recruit "opposition scientists" who speak out against mainstream climate science and portray the internal, scientific debate as ongoing (Ceccarelli 2011: 198). Cherry-picking is a technique that isolates the few experts in the minority of the scientific community that agree with a particular opinion, against accepted mainstream conclusions. Cherry-picking, then, subverts traditional understandings of authority as validated and supported by a larger, expert community. For example, the AI frequently cites Roy Spencer, "senior scientist at NASA's Marshall Space Flight Center" and known climate change skeptic (Beisner *et al.* n.d.: para. 60). Spencer is a legitimate scientist but is also a member of a slim minority of practicing scientists that still doubt anthropogenic global warming. The AI provides scientific legitimacy for its stalling of environmental action through appealing to the margins, such as Spencer. When one spends time on climate skeptical websites and in their communities, one starts to see the same names circulated as the true experts within a sea of other climate scientists. Membership in the scientific community affords credibility to the minority viewpoints of AI's cherry-picked skeptical scientists, whom the AI champions as the true climate scientists against the reigning scientific dogma.

In cherry-picking experts, the AI firmly sets itself in the minority of scientists with comparatively little support in the scientific communities for its conclusions. This strategy's success may be limited in scope, however, when the AI is compared to the larger and more widely recognized scientific bodies. At the very least, the bargainer strategy of highlighting of minority voices creates a stark dent in the argument that scientists have reached a "consensus" on climate change. While public understanding of the term often denotes 100 percent agreement, technical and scientific communities use the term to indicate extremely high probability of certainty. The appearance of the presence of any dissenting scientists may be interpreted by the public as mainstream scientists failing to account for all voices or simply exaggerating that a consensus exists at all. By cherry-picking evidence and borrowing from the voices of skeptical scientists, bargainers can appear to argue from within that community. Even a minority voice that has the veneer of scientific authority can pose a powerful threat to the public's understanding of science. By appealing to the minority of legitimate scientific voices that agree with them and adopting scientific standards that allow for alternate viewpoints, bargainers create the conditions for scientific revolution.

Bargainers thus create a double bind where climate scientists engaging in conversation are implicitly admitting that there is still a conversation to be had and climate scientists not engaging in conversation are overtly admitting that they have no evidence to shift skeptics from their position. In talking about the variety of topics from which he differed from mainstream topics, Victor argued that he was not sure if his beliefs "are mainstream or not." This comment indicates that this skeptic's identity was constructed contra alternative perspectives (Lake 1997), but that he was not certain whether his skeptical views were truly skeptical any more, or if they had become the new mainstream. In other words, the skeptic identified themselves as having a natural skepticism toward "a great many things," but that skepticism itself does not have to be outside of the mainstream. Because skepticism is naturally a part of scientific inquiry, laying claim to a skeptical identity is a strategy for presenting themselves as legitimate scientific voices (Bloomfield and Tillery 2019).

Part of relying on cherry-picked data seems to be the tendency to overgeneralize and make hasty generalizations about groups of people from limited examples. A few bargainers I spoke to reflected skepticism of mainstream scientific and academic authorities and preferred to do their own research to get the full picture of the current status of climate science. Abigail specifically referenced the consensus studies on climate change and denounced them as repeatedly disproven by other reports. Mostly citing newspaper editorials, Abigail saw the original articles as irrevocably corrupted and untrustworthy due to commentaries that questioned their methodology. When asked if she felt these problems with the consensus paper's peer review were universal, she agreed and noted that academics have to publish "to fulfill degree or publishing requirements," so are sometimes "sloppy or lazy" in the work they do. Because Abigail perceived that the scientific community is largely focused on publishing articles that agree with mainstream perspectives on anthropogenic climate change, she attributes the peer review process as containing no motivation or incentive to correct faulty work. In other words, the agreeability of the conclusion overshadows methodological or theoretical concerns to the detriment of scientific standards. Abigail elaborated, "it's easier [for academics] to go along with what everyone else seems to be saying," because "dissenting opinions" can be "dismissed easily," thereby damaging a scientist's publishing record. She also noted that the publication of reports that support anthropogenic climate change partially stem from "drawing conclusions based on using the same data set," and not being open to alternative data.

In addition to casting doubt on mainstream climate scientists by highlighting those voices who express minority viewpoints, bargainers also cast doubt on scientifically-sanctioned institutions such as peer review. While scholars have defined peer review as a communal process of certifying knowledge (Gross 1996; Prelli, 1989), bargainers view peer review as a label that scientists hide behind, certifying nothing but that a majority of scientists have been duped into believing the severity of climate change. Ceccarelli (2011) identified doubt in peer review as a common feature of climate skepticism rhetoric (and other unorthodox scientific perspectives). Targeting the process of peer review is persuasive, in part, because it undermines specific scientists and academics without casting doubt on the scientific process as a whole, leaving mercenary scientists and their reputations intact. Scientific institutions are thus in need of renovation and restructuring, but climate science itself is still upheld as an important way of knowing about the environment, albeit when viewed through a Biblical worldview.

The strategy of cherry-picking does not necessarily need to change people's minds fully to be successful. Indeed, the bargainers' position is particularly persuasive because they do not need to convince an audience to take a certain action, but simply to pause, consider alternatives, or question the environmentalists' proposed path forward. Pennock (2003) called this the "wedge strategy," also used by creationists, in which one side does not need to offer evidence for their own position, but simply enough counter-evidence to the mainstream scientific explanation to create uncertainty, doubt, and the need for more research. If media are constantly giving air time to oppositional scientists and bringing up minority opinions as equal counterparts to mainstream science, then the conversation will forever remain in the minds of the public over *if* climate change is anthropogenic and severe as opposed to *how* can we mitigate and solve it (Boykoff and Boykoff 2004). In this sense, bargainers do not seek to eliminate climate science or cleave their Christian identity from climate change, but they do seek to promote inaction, delay, and caution.

In a society that largely turns to scientific authority on matters of the environment,[1] bargainers make a convincing case that we should thus listen to all scientists and their claims. When evidence emerges that seems to contradict or is interpreted as contradicting mainstream science, bargainers use that evidence to call for a rethinking of climate science and its models, thus attempting to instigate a new scientific revolution. Furthermore, bargainers appeal to democratic ideals of open conversation to foster uncertainty, confusion, and doubt on mainstream scientists' opinions and convince audiences that perhaps there should be a new regime that takes into account alternative scientific viewpoints and opinions.

Conclusion

The discourse of bargainers shares rhetorical patterns that stretch standards of climate science, and science in general, to blend with Christian values. Oftentimes, bargainers strategically substitute economics as a proxy for their religious views to enter conversations and anchor their climate skepticism in a more accessible and communal concern. As bargainers, they appeal to open debate and view themselves as revolutionists hoping to alter mainstream understanding of climate science from within. Bargainers engage in a tragicomic frame which predicts mainstream science will eventually correct their misconceptions after figureheads of the movement are debunked and corrected. After this paradigm shift where previous conceptions and interpretations of climate science are refuted, true climate scientists, who are skeptical of the danger and severity of climate change, will become the reigning standard. As an exemplar bargainer, the AI's discourse challenges environmentally-focused science as the sole resource on which to base policy and offers a blended religious and economic framework from which to argue. Driven by a contemporary Protestant ethic that turns to financial autonomy as a sign of spiritual piety, bargainers use more varied argumentative resources than do the separators. While the separators eschew climate science and make enemies of scientific institutions, the bargainers shield themselves from critique that they perceive to be rational, well-researched, and legitimate contributions to climate deliberation. From their minority position, the bargainers see themselves as David appealing to the few others who share their opinion against the Goliath of other environmental groups, cherry-picking data, interpretations, and champions who continue the conversation long after the technical community has reached consensus.

Bargainers have discursive similarities with the separators and the harmonizers, but also differ from them in stark ways, making them a sort of middle ground among the categories. They share with the separators the overt climate change denial and ultimately seek to oppose environmental action. They share with the harmonizers a less aggressive, more comic frame and the view that climate science is valuable, can provide insight into reality (at least when tempered by religion). But, bargainers uniquely portray skepticism of environmental science's conclusions within a tragicomic narrative of revolution. This narrative identifies people and institutions as obstacles to the Christian, economic truth about climate change, but does not call for their elimination or sacrifice. Instead, most are simply to be set right, refuted through dialogue, and out-reasoned with competing interpretations.

As a coping mechanism against the rise of science, bargainers create an organizing framework that minimizes science's legitimate challenges to their Christian loyalties. Burke (1984: 23) argued, "As a given historical frame nears the point of cracking, strained by the rise of new factors it had not originally taken into account, its adherents employ its genius casuistically to extend it as far as possible." To blend mainstream climate science to their religious worldviews, bargainers appear to have stretched science near the point of becoming something wholly unrecognizable. Although they appear to appeal to scientific standards, scholars have pointed out such positions' similarities with archaic notions of science that do not align with how contemporary science is performed, but often aligns with public expectations of science (Pennock 2003). Instead of stretching Christian doctrine and hermeneutics to fit climate science, as we will see the harmonizers do, the bargainers undergo great epistemic stretching and "considerable enterprise" to make room for Christianity in the reigning scientific interpretations of the environment and our current climate crisis (Burke 1984: 184). This stretching is a difficult process, but it is a simpler, more comforting process than abandoning previous frames (e.g., Christianity or neoliberal economics) when they appear to conflict with climate science.

While bargainers are, typically, more open to discussion and engagement with mainstream climate science than separators it does not mean that conversations with them are easy or straightforward. Because bargainers have their own scientific authorities, substitute cherry-picked data, and hold different priorities (such as economics) over the environment, talking to bargainers is oftentimes a complicated process. As evidenced by my conversations above, bargainers have often done extensive research to support their perspective and seek out alternative resources to support their arguments and poke holes in mainstream interpretations of scientific data. The bargainers' performance of climate skepticism constructs a worldview where revolution and challenging mainstream climate science contribute to productive conversation, instead of detract from it. Attempts to redirect the conversation to the "real" climate science or to denigrate the bargainers' cherry-picked evidence are likely to be met with claims of unscientific silencing.

Many of the bargainers I spoke with seemed entirely sincere in their beliefs. They were concerned with people corrupting the scientific process, hiding information from the public, and borrowing the veneer of science to further their own goals. In accepting their sincerity and listening to those concerns, I aimed to learn more about bargainers and locate where they might find overlap with environmentalism. The next chapter outlines various strategies to engage bargainers, listen to and understand their perspective, and avoid rhetorical traps that bargainers (intentionally or unintentionally) might set to characterize environmentalists as close-minded and unscientific. Similar to strategies for engaging separators, it is important to recognize the values and framework of bargainers to have open, welcoming conversations. We must also consider their discursive differences and distinct starting frameworks to respond in ways that take advantage of theoretical conclusions and tailored approaches.

Note

1 There are competing discourses about the faith that society has in scientific institutions. It seems that for some people on certain topics, science is still an ultimate authority, but for others there is significant doubt and uncertainty. See Goodwin and Dahlstrom (2014) for an extended discussion.

References

Abigail [pseudonym]. (2018) personal communication [digital exchange].

Acton Institute (AI) (n.d.-a) Acton University. Available at: http://university.acton.org/about-au (accessed August 7, 2018).

Acton Institute (AI) (n.d.-b) History of Acton Institute. Available at: https://acton.org/about/history-acton-institute (accessed August 7, 2018).

Acton Institute (AI) (n.d.-c) Our Mission & Core Principles. Available at: https://acton.org/about/mission (accessed May 10, 2018).

Anderegg, W.R.L., Prall, J.W., Harold, J., and Schneider, S.H. (2010) Expert Credibility in Climate Change. *Proceedings of the National Academy of Sciences* 107(27): 12107–12109. DOI: 10.1073/pnas.1003187107.

Ballor, J.J. (2006) Stewardship and Economics: Two Sides of the Same Coin. Acton Institute, February 15. Available at: https://acton.org/node/3671 (accessed August 7, 2018).

Ballor, J.J. (2015) Christian Reformed Church Backs UN Climate Change Agenda. Acton Institute, July 8. Available at: https://acton.org/pub/commentary/2015/07/08/christian-reformed-church-backs-un-climate-change-agenda (accessed August 7, 2018).

Beers, J.M., Hittinger, R., Lamb, M., Neuhaus, R.J., Royal, R., and Sirico, R.A. (n.d.) The Catholic Church and Stewardship of Creation. Acton Institute. Available at: https://acton.org/public-policy/environmental-stewardship/theology-e/catholic-church-and-stewardship-creation (accessed May 10, 2018).

Beisner, E.C., Cromartie, M., Sieger Derr, T., Knippers, D., Hill, P.J., and Terrell, T. (n.d.) A Biblical Perspective on Environmental Stewardship. Acton Institute. Available at: https://acton.org/public-policy/environmental-stewardship/theology-e/biblical-perspective-environmental-stewardship (accessed August 7, 2018).

Bloomfield, E.F. and Lake, R.A. (2015) Negotiating the End of the World in Climate Change Rhetoric: Climate Skepticism, Science, and Arguments. In: Meisner, M.S., Sriskandarajah, N., and Depoe, S.P. (eds.) *Communication for the Commons: Revisiting Participation and the Environment*, pp. 384–396. Uppsala, Sweden: The International Environmental Communication Association.

Bloomfield, E.F. and Tillery, D. (2019) The Circulation of Climate Change Denial Online: Rhetorical and Networking Strategies on Facebook. *Environmental Communication* 13(1): 23–34. DOI: 10.1080/17524032.2018.1527378.

Boykoff, M.T. and Boykoff, J.M. (2004) Balance as Bias: Global Warming and the US Prestige Press. *Global Environmental Change* 14(2): 125–136. DOI: 10.1016/j.gloenvcha.2003.10.001.

Bray, D. (2010) The Scientific Consensus of Climate Change Revisited. *Environmental Science & Policy* 13(5): 340–350. DOI: 10.1016/j.envsci.2010.04.001.

Burke, K. (1984) *Attitudes toward History*. Berkeley, CA: University of California Press.

Carlson, A.C. (1986) Gandhi and the Comic Frame: "Ad Bellum Purificandum." *Quarterly Journal of Speech* 72(4): 446–455. DOI: 10.1080/00335638609383787.

Ceccarelli, L. (2011) Manufactured Scientific Controversy: Science, Rhetoric, and Public Debate. *Rhetoric Public Affairs* 14(2): 195–228.

Channell, J., Jansen, H.R., Syme, A.R., Morse, E.L., Savvantidou, S., and Yuen, A. (2013) Energy Darwinism. Citi GPS: Global Perspectives & Solutions, Citigroup, October 15. Available at: www.citivelocity.com/citigps/ReportSeries.action?recordId=21 (accessed May 29, 2018).

Channell, J., Jansen, H.R., Syme, A.R., Nguyen, P., Prior, E., Curmi, E., Rahbari, E., Morse, E.L., Kleinman, S.M., and Kruger, T. (2015) Energy Darwinism II. Citi GPS: Global Perspectives & Solutions, Citigroup, August 14. Available at: www.citivelocity.com/citigps/ReportSeries.action?recordId=41 (accessed May 29, 2018).

Christiansen, A.E. and Hanson, J.J. (1996) Comedy as Cure for Tragedy: Act Up and the Rhetoric of Aids. *Quarterly Journal of Speech* 82(2): 157–170. DOI: 10.1080/00335639609384148.

Cook, J., Nuccitelli, D., Green, S.A., Winkler, B., Painting, R., Skuce, A., Richardson, M., Way, R., and Jacobs, P. (2013) Quantifying the Consensus on Anthropogenic Global Warming in the Scientific Literature. *Environmental Research Letters* 8(2): 024024. DOI: 10.1088/1748-9326/8/2/024024.

de Onís, K.M. (2012) "Looking Both Ways": Metaphor and the Rhetorical Alignment of Intersectional Climate Justice and Reproductive Justice Concerns. *Environmental Communication* 6(3): 308–327. DOI: 10.1080/17524032.2012.690092.

Desilet, G. and Appel, E.C. (2011) Choosing a Rhetoric of the Enemy: Kenneth Burke's Comic Frame, Warrantable Outrage, and the Problem of Scapegoating. *Rhetoric Society Quarterly* 41(4): 340–362. DOI: 10.1080/02773945.2011.596177.

Dixon, G.N. and Clarke, C.E. (2013) Heightening Uncertainty around Certain Science: Media Coverage, False Balance, and the Autism-Vaccine Controversy. *Science Communication* 35(3): 358–382. DOI: 10.1177/1075547012458290.

Doran, P.T. and Zimmerman, M.K. (2009) Examining the Scientific Consensus on Climate Change. *Eos, Transactions American Geophysical Union* 90(3): 22–23. DOI: 10.1029/2009EO030002.

Fahnestock, J. (1986) Accommodating Science: The Rhetorical Life of Scientific Facts. *Written Communication* 3(3): 275–296. DOI: 10.1177/0741088386003003001.

Friends of Science Society (2014) Friends of Science. Available at: www.friendsofscience.org/index.php?id=2315 (accessed March 19, 2018).

Goodwin, J. and Dahlstrom, M.F. (2014) Communication Strategies for Earning Trust in Climate Change Debates. *Wiley Interdisciplinary Reviews: Climate Change* 5(1): 151–160. DOI: 10.1002/wcc.262.

Gross, A.G. (ed.) (1996) *Rhetorical Hermeneutics: Invention and Interpretation in the Age of Science*. Albany, NY: State University of New York Press.

Jensen, G. (2015) Nature, Markets, and Human Creativity. Acton Institute, May 27. Available at: https://acton.org/pub/commentary/2015/05/27/nature-markets-and-human-creativity (accessed August 7, 2018).

Joseph [pseudonym]. (2018) personal communication [digital exchange].

Kuhn, T.S. (1996) *The Structure of Scientific Revolutions*. Third edition. Chicago, IL: University of Chicago Press.

Lake, R.A. (1997) Argumentation and Self: The Enactment of Identity in *Dances With Wolves*. *Argumentation & Advocacy* 34(2): 66–89.

Matthews, P. (2015) Why Are People Skeptical about Climate Change? Some Insights from Blog Comments. *Environmental Communication* 9(2): 153–168. DOI: 10.1080/17524032.2014.999694.

McCright, A.M. and Dunlap, R.E. (2003) Defeating Kyoto: The Conservative Movement's Impact on US Climate Change Policy. *Social Problems* 50(3): 348–373.

McGee, M.C. (1990) Text, Context, and the Fragmentation of Contemporary Culture. *Western Journal of Speech Communication* 54(3): 274–289. DOI: 10.1080/105703 19009374343.

Nothstine, R. (2015) After the Culture Wars. Acton Institute, November 11. Available at: https://acton.org/after-culture-wars (accessed August 7, 2018).

Pennock, R.T. (2003) Creationism and Intelligent Design. *Annual Review of Genomics and Human Genetics* 4(1): 143–163.

Phillips, B.B. (2010) Evangelicals and Global Warming. Acton Institute, June 23. Available at: https://acton.org/pub/commentary/2010/06/23/evangelicals-and-global-warming (accessed August 7, 2018).

Prelli, L.J. (1989) The Rhetorical Construction of Scientific Ethos. *Evolution* 34(5): U980.

Schwarze, S. (2006) Environmental Melodrama. *Quarterly Journal of Speech* 92(3): 239–261. DOI: 10.1080/00335630600938609.

Shaftel, H., Jackson, R., and Callery, S. (2018) Scientific Consensus: Earth's Climate Is Warming. NASA: Climate Change and Global Warming. Available at: https://climate.nasa.gov/scientific-consensus (accessed August 7, 2018).

Smith, F.M. and Hollihan, T.A. (2014) "Out of Chaos Breathes Creation": Human Agency, Mental Illness, and Conservative Arguments Locating Responsibility for the Tucson Massacre. *Rhetoric and Public Affairs* 17(4): 585–618. DOI: 10.14321/rhet publaffa.17.4.0585.

Snow, C. (2015) Unholy Alliance: Who Is Advising Pope Francis on Global Warming? Acton Institute, June 17. Available at: https://acton.org/pub/commentary/2015/06/17/unholy-alliance-who-advising-pope-francis-global-warming (accessed August 7, 2018).

Victor [pseudonym]. (2018) personal communication [digital exchange].

Walsh, L. (2013) *Scientists as Prophets: A Rhetorical Genealogy*. Oxford; New York: Oxford University Press.

4 Bargainer strategies

Working within frames, joining the revolution, and employing examples

Learning the differences between bargainers and separators is integral to understanding how "multifaceted, complex, and nuanced opposition" to mainstream climate science is (Bloomfield and Tillery 2019: 32). While the separators and bargainers are not always discrete, exclusive categories, my labeling them as such works to identify various constellations of values, priorities, and perspectives that comprise different community identities and their guiding frameworks within climate denial communities. These two groups represent skeptical responses to climate change rooted in religious perspectives. But, that does not mean that everyone who expresses these views is religious nor that an individual will solely use strategies from a single category. These exemplars should be used as heuristics to unpack the variety of skeptical perspectives and how religion influences particular interpretations. These differences are also important because they guide how we might engage and respond differently to skeptics' rhetorical strategies. In other words, while separators and bargainers are both contained under the label of "climate skeptics," they perform and enact their skeptical, religious identity in ways that prompt different approaches and responses.

A primary difference previously outlined between the separators and bargainers are their guiding metaphors. The separators' war metaphor produces more defensive discourse, whereas the bargainers' metaphor of revolution leads to discourse that emphasizes climate science's uncertainty, corruption, and the need for more research. This discourse may appear reasonable through the bargainers' worldview of scientific revolution, but it ultimately serves to delay productive conversations and promote disingenuous controversy (Ceccarelli 2011). Unlike the aggression of the separators, the biggest rhetorical trap of the bargainers is their perceived neutrality and dedication to scientific standards. These features of bargainers' discourse serve to deflect conversations from the reality of climate change. Instead of challenging authority, bargainers cherry-pick data and find "contrarian" scientists to advocate for their perspective (Dunlap 2013: 693). Instead of denying climate science as an authority, bargainers appeal to scientific standards when countering environmental information. When engaging bargainers, the issue, then, is not a matter of avoiding scientific authorities (as is with the separators), but shifting their perspective on whose scientific voice matters and how disagreement among scientists does not always indicate that a revolution is on the horizon (Ceccarelli 2011). Based on my interactions with people who exhibit bargainer characteristics, I found that bargainers are usually willing to engage in discussion, have done extensive research into certain areas of climate science or the work of particular climate scientists, and are ready with counterarguments and refutations. For example, Abigail sent me a link to a few articles she had read about "how to refute climate skeptics" and found them wholly unconvincing.

Considering the rhetorical features of the bargainers (i.e., constructing a Biblical filter, appealing to scientific standards, and cherry-picking), I offer three strategies to engage them in conversation and to shift their understanding of scientific revolutions and proper ways to interpret climate science. These suggestions are based on my analysis of exemplar discourse of climate skeptic groups and in-depth conversations with people who exhibited bargainer characteristics. The three strategies proposed here stem from the book's dialogue and rhetorical listening perspective and focus on understanding and engagement over coercion or overt persuasion (Johannesen 1974; Ratcliffe 1999). The three strategies are working within frames (namely the bargainers' blended Biblical, economic frame), joining the revolution, and employing examples. These strategies lean in to the bargainers' frequent willingness to engage conversations while altering frameworks and assumptions from within.

George Lakoff (2010) argued that altering vocabularies within frames is incredibly difficult but that it can be a successful strategy as long as there is enough alignment between the original frame and the new vocabulary. Kenneth Burke (1984: 184) would agree with such assertions, similarly arguing that people are reluctant to change their perspectives and will take on "considerable mental enterprise" in adjusting their original frame before abandoning it. To respond to the bargainers' stretching of climate science to fit their Christian worldview, we must work within their existing frameworks and offer enough counter-examples, new evidence, and altered vocabulary to prompt reconsideration. With some continued conversation and exchange of information, we can "introduce new principles" of pro-environmental attitudes and behaviors while allowing them to stay "faithful to old principles" of scientific revolution and skepticism (Burke 1984: 229). Neither separators nor bargainers will change their minds in a single conversation; it is up to climate communicators and environmentalists to provide resources, prompt questioning and critical thinking, and sow the seeds for future attitude and behavior change. This will only occur through disrupting existing associations and frames so that down the line the altered frame can lead to more environmentally-restorative attitudes and behaviors. In what follows, I describe the theoretical justification for these strategies and detail interactions from multiple conversations with bargainers that demonstrate the application of these strategies.

Working within frames

When addressing separators, we discussed the strategy of asking questions and accepting premises in order to reason from shared values and beliefs to environmentally-friendly attitudes. For bargainers, we want to follow a similar argument chain by which we frame and modify our arguments to fit bargainers' values. But, this strategy is tailored to bargainers by identifying and adjusting frames instead of specific premises. The exemplar bargainer, the Acton Institute (AI), and the bargainers I spoke to often referred to the economic relationship between personal autonomy, faith, and climate change. Bargainers, operating under Christian and economic frames, tend to legitimize grounds and evidence based on their support of economically-friendly policies, personal autonomy, and a concern for *human* life. In my conversation with Victor, he noted that he came "from a low-income background" so primarily viewed environmental policies in terms of economic "burden on poor households." The frame through which this bargainer understands climate change is with economics in mind. His primary concern was for policies to not be "too burdensome" on economic success or economic autonomy. Abigail shared this economic framing, arguing in an online chat exchange, "In a free market capitalist society, it all boils down to $." No matter the discussion to be had in other areas, these bargainers brought the conversation back to economic autonomy and profit.

The bargainers I spoke with shared a common affinity for industrialization and the comforts that it has provided. Abigail noted that many environmentalist arguments to minimize consumption and live more simply are "naïve" and fail to acknowledge all that industrialization has provided for humanity. This bargainer accused environmentalists of distorting the original Genesis story, substituting Eden with pre-industrial life and "the Fall" with the development of agriculture and industry. In this bargainer's story, it is nuclear power and "real competition among innovators" in industry and technology that will actually save us. In response to this line of dialogue, I asked if the narrative could include both industry and limited consumption; in other words, is it only impossible to follow the values of environmentalists and minimalists at the expense of industry or is there a way for them both to exist. Abigail responded that she considered a primary problem with environmentalism is that climate change will "always end up being an over-population problem." She believed that following this line of thinking will result in "inevitably killing large groups of low-influence people." Under the bargainers' framework, even uniting industry and environmentalism will boil down to the loss of human life and costs in the "millions of dollars and jobs."

Once we understand the guiding frames of our bargainers, which were primarily personal autonomy (guided by Christian beliefs) and capitalist profit, we can respond to bargainers by meeting them where they already are. While there are risks associated with embracing a neoliberal ideology to talk about the environment, there are also potential opportunities for engagement in working within frames. At the very least, working within frames is much less taxing than trying to move the conversation to a different frame as such moves might also be viewed negatively by our dialogue partners. By staying within the frames bargainers value, we can maintain their attention and interest in continuing discussion without being perceived as dismissing their frames out of hand. There is existing research that supports the use of "broker" categories that mediate between held values and the environment (Hoffman 2011: 3). Interestingly, Hoffman (2011) identifies religion as one of these broker categories, which is certainly the case for harmonizers. For bargainers (and separators), however, meeting on topics of faith is complicated because of their specific definitions and interpretations. Instead, we could turn to Hoffman's (2011: 3) broker categories of technology and security or attempt to "redefine[e]" the relationship between economics and the environment. In my conversations with people across the three categories, there was near universal support for investing in new techno-logy and protecting public health. Using these accepted frames couches environ-mental benefits in more concrete, personal terms that more directly affect people such as "economic growth, better healthcare, and more secure jobs" (Elliott 2014: 244). These strategies for engagement coalesce around the idea of finding common ground and shared substance where bargainers already have interest, see value, and view as important in order to lead to mutual understanding and potentially persuasion.

Bargainers, who use Biblical and economic filters to reach environmental conclusions, may conflate environmentalism with the complete eschewing of economics and industry. When discussing potential environmental policies, Victor lauded the replacement of fossil fuels with other power sources (namely nuclear), funding more public transportation, investing in electric vehicles, and restricting air pollution. Abigail argued that "science" is going to have to solve our current disputes over climate change, referring to technological advance-ments and adaptation. She continued that "if sustainable and renewable power is the solution, I really don't care where the electricity comes from as my com-puters and phones and tablets … work." Taking these comments into considera-tion, we might note that bargainers are open to taking environmentally-friendly actions (and that I did little convincing to get them there), but that those behav-iors need to be brought out of an environmental frame and emphasized as com-patible with faith, the economy, and personal comfort. Indeed, bargainers expressed what could be considered pro-environmental attitudes, but differed on the extent to which they prioritized these policy initiatives and to which they saw them as part of environmentalism.

For example, in exchanging articles about renewable energy, Lucas argued that renewables could play an important role in the future but could not "alone" provide enough energy. In this sense, Lucas already held some ideas that were compatible with environmental thinking. This interaction is important because it challenges the assumption that being a skeptic means to doubt or deny climate change itself or the factors we argue contribute to it. No one I spoke to in gather-ing my research denied that temperatures were increasing, but the severity, importance, or urgency of reacting to that information was disputed. Knowledge campaigns and attempts to correct information deficits, therefore, may fail, because it is more how the bargainers are interpreting the facts than not knowing the facts. A strategy for dealing with bargainers that works within their frame-work is not trying to insist that the conversation must end with an assertion of being an environmentalist for the conversation to have been successful. For example, Victor surprised me by asserting that he does care about climate change and the environment, but "worr[ies] about other things more." Victor felt that he did not align with mainstream climate change attitudes because his ranking of priorities was different, not because he did not care or value the environment at all.

Joining the revolution

It is dangerous and damaging to engage the guiding metaphor of the separators because the war framework immediately places one on the side of the enemy. In rejecting their framework, it is still likely they will view conversation partners as enemies, but the risk is minimized by skirting the notions of war, opposition, and competition. The bargainers' metaphor, while still positioning themselves as revolutionary forces against mainstream climate science, is not a framework that should be avoided. The key aspect of my suggestion to join the revolution is not to switch sides, as it were, and begin to question climate change. Instead, this suggestion is meant to encourage us to operate from within the bargainers' worldview, which does share the common value of science and the discovery of the strongest, most valid science. In other words, we must construct a competing interpretation of the scientific revolution so that science has already undergone a revolution (in recognizing anthropogenic climate change) and is not currently undergoing another.

Shifting bargainers' perspectives on revolution involves turning the bargainers' critiques of mainstream science against climate skepticism and holding their arguments to logical and scientific standards. Taylor (1992) argued that when scientists labeled intelligence design and creation science as religious and non-scientific, they did themselves a disservice. Instead of measuring intelligent design against scientific standards and thus labeling it a weak science, the scientific community has unwittingly "allowed [creationism] to escape the rigorous empirical tests to which scientific claims are traditionally subjected" (Taylor 1992: 285). The same argument can be made in the case of climate change. To, by default, disqualify climate skeptics from the conversation because they are not rational or scientific, we miss the opportunity to evaluate, engage, and measure the skeptics' science against mainstream climate science. Empowering our audience to evaluate both our messages and their own can frame the conversation as a mutually beneficial knowledge exploring, with the hope of opening up more opportunities for trust (Goodwin and Dahlstrom 2014). If we, instead of dismissing naysayers, acknowledge their presence as part of the scientific process, we invite participation in their revolutionary narrative. This rhetorical move situates us as listeners and dialogue partners (Johannesen 1974) operating under a similar perspective interested in evaluating and moving forward with the best climate science without blindly accepting authority. While there are certainly risks involved in such rhetorical moves, the alternative is to fail to engage and ignore bargainers as an audience in the climate change controversy. Furthermore, opening ourselves up to risks can be an important step in establishing trust and good will (Goodwin and Dahlstrom 2014).

In sympathizing with the revolutionary identity and the power of skepticism, we can find starting points for productive deliberation. Once we have established that we ourselves are concerned with scientific authority and the possibility of revolution, we can then work within that framework to question the bargainers' premises for how science works, offer alternative perspectives on how revolutions occur, and question if climate skepticism qualifies as a scientific revolution. This strategy also makes use of the idea of "inoculation," where people are exposed to likely counter-arguments in an attempt to prevent their acceptance later on (Moran *et al.* 2016). For example, we might discuss with bargainers the frequent arguments against climate scientists that they are engaged in conspiracy or are being paid to fabricate results. In offering a weakened, exaggerated version of those common arguments, the hope is that our interlocutors will be prepared to dismiss them or will be primed to question them when they encounter them. This strategy encourages us to address arguments up front and acknowledge their existence but dismiss them as talking points instead of legitimate arguments. In making known the strategies of bargainers and the common arguments against mainstream climate science, we can prompt the bargainers' skeptical and critical mindset to evaluate skeptical arguments instead of ones in support of climate change.

Taylor (1992: 279) argued that many science deniers take a Baconian view of science, which posits tenets of falsifiability that privilege "close empirical observation and strict processes of induction from those observations." This perspective mirrors public understanding of how science operates, even though this is not an accurate view of how science typically operates (Taylor, 1992). While Taylor (1992) focuses on the creationism and evolution debate, the same tactics are used in the climate change controversy, where climate scientist's predictions into the (yet unobserved) future are compared to fortune teller's prophecies (Bloomfield and Lake 2015). One bargainer I spoke with, Steven, viewed climate scientists as being arrogant when making predictions about future temperature increases that are not rightly earned by current research. He further argued that it is near impossible to model a complex, chaotic system such as climate change and that alone was a reason to doubt climate change science's conclusions. Although there are many potential risks in appealing to predictive science (e.g., Bloomfield and Lake 2015; Walsh 2013), it is important in conversations with bargainers to emphasize the power of climate models, while acknowledging their limitations, and the success they have had in predicting changes in temperature, sea level rise, arctic melt, and other climate change markers. A failure to acknowledge these inherent probabilities and uncertainties may contribute to skepticism of those models.

Ceccarelli (2011: 198) argued that climate deniers (whom I would label bargainers) embrace a narrative that portrays them as "heroes in an unfolding scientific revolution." To try and deny this narrative only provides further validation for it. Instead of telling bargainers that they are not heroes or are not on the right side, we can instead reframe the presence of revolutionary thinking as a natural occurrence. In other words, dissent, uncertainty, and contrarians are not markers of an impending revolution, but an ever-present characteristic of science. In embracing the framework of revolution, we should correct the bargainers' assumption that the presence of naysayers indicates an imminent shift. We should not pretend that science is always right or not subject to change, but that also does not mean that science always functions with 100 percent consensus. Consensus, as defined in public understanding, denotes complete agreement. A scientific consensus, however, is based on probabilities and likelihoods of causal relationships. To deny the inherent uncertainty in science provides skeptics with more ammunition and validates counter-examples as paradigm-shattering.

We can join the revolution by "retell[ing] the Kuhnian story" to include that the "mere existence of dissent does not mean that a scientific revolution is underway" (Ceccarelli 2011: 215). Instead, we can tell a story of the revolutionary features of climate denial as remnants of the "old paradigm" that was unable to convince the scientific community that new thinking was needed. In my interviews with bargainers, it was clear that many of them held Baconian views of science and were concerned about the continued presence of skeptical scientists. But, the Baconian view of science is not always relevant in current scientific endeavors and climate science is no exception. These premises and preconceived notions exist, so tackling them will help to unhook bargainers from "illusory correlations" between their assumptions and views of climate science (McFadden 2016). Theoretically grounded in an understanding of bargainers' revolutionary metaphor and Ceccarelli's (2011) suggestions for countering revolutionary ideas, I attempted to convince a bargainer of the importance of climate science despite the manufactured appearance of a lack of consensus (Ceccarelli 2011).

In my continued conversion with Victor, he noted that there had been much "suppression of dissent" in climate science, making the consensus seem more powerful and robust than it actually was. He also described his own dismay at climate scientists not knowing how "to engage" climate skeptics and respect their alternative viewpoints. Victor also accused climate scientists and the media of exaggerating the current situation for their own personal gain. He particularly noted the "baggage" that comes along with claiming to be an environmentalist, climate scientist, or academic, because prominent ones (such as John Cook, Michael Mann and Stephan Lewandowsky) had stirred up much "distrust" in the skeptical community. In order to highlight that the mere presence of detractors does not necessarily mean that the science is wrong, I brought up the anti-vaccine community and the Flat Earth community. I asked Victor if the presence of people who deny the medical benefit of vaccines and that the Earth is a globe indicate revolutions in these areas. I chose these examples instead of creationism (which is another community that I study) to avoid defensiveness from addressing an issue with a prominent Christian component. Victor immediately responded that there was far more evidence on the side of climate skepticism than for anti-vaxxers and Flat Earthers. He also noted that the discussion over the shape of the Earth has very limited consequences, while decisions about climate change can have incredibly important impacts on people's freedom and financial wellbeing. I pushed back by emphasizing the anti-vaccine example and by highlighting the similarities between the potential health impacts of anti-vaccine discourse and climate skeptic discourse. Victor insisted that the situations were different but stopped offering additional reasons for those perceived differences. I concluded this part of our conversation by reiterating that the mere presence of deniers does not always mean that a shift in scientific thinking is on the horizon; sometimes, people disagree with mainstream science without being unfairly silenced.

Victor later noted in our conversation that he was not sure about the exact numbers, but he believed the 97 percent consensus statistic is a fabrication because of how many skeptics he interacts with. He argued that "if" the weight of evidence for climate change was truly astonishing and without reproach, then surely there would not be a single defector. To respond to this argument, I prompted Victor to reflect on the nature of his perceived revolutionaries. I prompted him to consider that there seem to be the same voices emerging again and again, publishing articles, appearing in the media, and promoting the lack of consensus. He responded that he "hasn't really noticed," but felt that any dissenters painted a poor picture for the accuracy of climate scientists' results. He argued that there are "plenty of references" about how climate change is a hoax, and that makes him skeptical of other conclusions. It was clear in my interaction that Victor consumed information in an echo chamber where he saw an increasing amount of information shedding doubt on the consensus and not as much, if any, information that confirmed it. Research confirms that news coverage of climate change may often emphasize the presence of dissent (Dixon and Clarke 2013), which can damage faith in scientists and their conclusions about climate change and inadvertently level the scales between scientists and skeptics (Brüggemann and Engesser 2017). Bargainers frequently sent me links to skeptical and denial websites, blogs, and Facebook groups, indicating that they mostly consume media and information that reinforces one another and separates its audience from actual climate science (Bloomfield and Tillery 2019). What they perceived to be a large, networked army of sources served as an argumentative resource to support their notion that climate skeptics were onto something (Paliewicz 2018).

While 3 percent is obviously dwarfed by 97 percent, the presence of dissent feeds into bargainers' narrative of a burgeoning revolution. To try and counter this "false balance" (Dixon and Clarke 2013) between climate science and climate skepticism, I encouraged Victor to get outside of his traditional media sources and explore others that may challenge their perspectives. If, as they say, science is about equality for various perspectives, then should he not give some time and attention to mainstream climate science? Victor agreed to read some of my material, so I started sending him one credible link from a validated, peer-reviewed source for every blog post or op-ed that he sent me. I think that the willingness to engage my material was fostered through my consumption of his sources where we traded resources, authorities, and links for additional information. Although I pointed out flaws and countered points made in the articles sent to me, my openness to read and discuss the sources built mutual respect and trust (Goodwin and Dahlstrom 2014) where Victor was open to reading mine.

Specifically, I provided additional articles and commentaries about the consensus reports that Victor themselves had not read but had only read skeptical critiques of. He said that he would explore the original articles. But, in his response, Victor noted that the consensus should "be closer to 100 percent, in that the climate is, and always has been, changing." When I asked for clarification on this statement (which appeared at first blush to be a startling logical contradiction), Victor noted that these studies simply asked scientists, "Is climate change real?" so had no real explanatory value. I again prompted him to read the articles, because they go into far more detail about the views of scientists surveyed and their resulting agreement on the severity of climate change and that it is human-caused. I clarified that when scientists use the term "climate change," they are not discussing general, natural changes in climate, but a specific, anthropogenic, and devastating altering of the Earth's natural systems. Again, Victor acknowledged that he did not know the original article's questions and would investigate them further. He seemed to recognize that they had made an error in his previous interpretations or at least was open to questioning those assumptions. In this sense, the bargainers' skeptical nature and the possibility for revolution was not something to dismiss, but a high standard to hold climate skepticism to. In other words, revolution is possible, but it is incredibly difficult, unlikely, and requires immense disruptions in established knowledge. For Victor, the phrase "climate change" may prompt a memory of our conversation and my correction on his interpretation of what "climate change" means in technical circles. Even if he did not accept my viewpoint right away, the repeated memory of the interaction and critical thinking about what it means that scientists agree that "climate change" is happening may likely have been altered.

The topic of revolution came up as well in my conversation with Abigail. We discussed how dissent is a natural part of the scientific process, where disagreements and debates spur new ideas and understandings. I noted that sometimes these debates lead to shifts in understandings that we might call "scientific revolutions," but these are few and far between. Abigail noted,

> I agree there are significant shifts as we fine-tune our understanding of God's creation. I've seen this shift through our understanding of the atom, aviation development, internal combustion, computers, programming, gaming, phones, just about everything we see and use every day.

I clarified her response by noting that revolutions often occur in our fundamental understanding of how the world works, not simply new technologies, but she was right that these shifts are often prominent and well-established shifts that cleanly produce a new paradigm from an old one. I mentioned that our previous state of knowledge was that humans did not influence the environment, but we had undergone a revolution in acknowledging our role in its now altered state. She responded that she "never really bought into" the shift to climate change from our original perspective, so does not see it as a revolution. We discussed further about the details of a scientific revolution and I left her with the notion that a scientific revolution has never been reversed. For example, scientists did not turn back to alchemy, reembrace experiments testing luminiferous ether, or begin to doubt gravity. There may still be scientists engaging these ideas, but dissent and alternative viewpoints are simply part of the social construction of scientific knowledge (Gross 1994).

When I reached out to Abigail a few months after our conversation, she noted that she did not have a "change of heart" on being a skeptic, but she did "appreciate the opportunity to look these things up" for more information. Although this particular conversation did not prompt immediate attitude shifts, I was not discouraged. I viewed my conversation with Abigail as successful because it was a sustained, genuine interaction where the bargainer was reading new material outside of her echo chamber and considering the nature of scientific revolutions. Throughout this book, I make no claims that single interactions can move people's entire perspectives and worldviews – this would be near impossible. Beliefs linked to one's faith are incredibly hard to shift and displace, especially by single interactions. Many people who have altered their views on climate change have done so through exploring the research on their own and coming to their own conclusions (Matthews 2015). This conversation and others I had with bargainers provide hope that they will try to seek out new outlets of information, trust their own interpretations instead of always turning to a skeptical source, and were able to hear counter-examples to challenge their staunch rejection of the consensus on climate change, academics, and what counted as reliable sources.

Employing examples

Because we know the frames within which bargainers work, we can turn to faith and economics for powerful, personal examples that bargainers will attend to. The strategy of employing examples comes directly from persuasion research that details that feelings of personal involvement will lead to increased persuasion. For example, Dillow and Weber (2016) argued that people who were given narratives about the need for organs in their state were more likely to sign up to be organ donors because they felt they might one day be in need. Although I did not directly ask bargainers about their political affiliation, there is a strong association between conservatism and climate change skepticism (Pew Research Center 2013). Given this association, it is also a persuasive strategy to connect with conservatives on concrete values and examples, since they share a stronger affinity for the concrete over the abstract (Bloomfield and Katula 2012). When asked what concerns they had about the environment, Abigail noted her concern about pollution, especially when it led to "toxins, birth defects, cancer rates, early deaths, etc." because of her commitment to "prevent[ing] further loss of life" and "ensur[ing] the safety of others." In response, I brought up that some Christian environmentalists consider taking care of the environment part of the pro-life movement because a healthy Earth would lead to healthy children. Abigail responded that if there were ways to both improve the environment and protect lives, "I probably wouldn't think twice about it." She emphasized that the primary goal of "dominion" is to be responsible for "prior/current/future generations while still living a fulfilling life." This exchange provides evidence that if we can package ideal environmental behaviors as examples of pro-life values without undue sacrifice, we can encourage environmentally-friendly behaviors without activating skeptical resistance.

Abigail noted that some of her skepticism comes from "when we see things like extra snow and colder temps" among other examples. For her, there was a connection between weather and climate that supported her perspective that there was not enough evidence for climate change. When I offered that the features she mentioned were associated with weather and not climate and that even extreme cold weather could be associated with increased greenhouse gases and climate change, Abigail noted that she would have to do more research. I brought up specific examples such as Death Valley having the hottest recorded temperatures on record this past year (Gabbatiss 2018) and that the ten hottest years the Earth has ever experienced have all happened since 1998 (Climate Central 2018). These counter-examples served two purposes. First, I sought to disrupt Abigail's association with specific weather events and climate by providing multi-year, aggregate information. Second, I sought to offer competing examples that point to overall increased temperatures despite individual instances of colder temperatures. As with other bargainers, Abigail was open to reading my resources and doing more personal research as our conversations progressed.

In addition to offering references and statistics to bargainers, we can also employ personal examples to engage them in conversation. The ability for us to employ examples about certain topics will greatly depend on who we are as communicators, how we came to care about the environment, and our personal histories. I was able to engage with bargainers as an academic and provide a prominent counter-example of their own biases against "elites." For example, when Lucas told me about the poor state of academic publishing and his perception that academics receive money for publishing, I offered my own experiences as refutation. Initially, I agreed with Lucas and lamented the "publish or perish" mindset that emphasizes frequent publishing. I then added that quantity is not always superior to quality, as one needs both to secure tenure at most institutions. I shared my personal journey through publishing and how rigorous the process of peer review is. I further noted that I am a public employee where my salary is open to view by anyone and that the only payments I get from my work are regular paychecks tied primarily to fulfilling teaching duties. While Lucas was initially derisive toward academics, his tone quickly changed when I asked if he felt that I was corrupt and perform bad research because I was an academic. Lucas did not engage in similarly harmful or derogatory generalizations and conceded that surely not *all* academics were forcing through low quality publications. He further noted, however, that even if I did not do "sloppy or lazy" work, some academics might. I see this exchange as a potential reworking of Lucas's preconceived ideas about academia. By offering myself as a counter-example, I positioned myself a sincere dialogue partner who had established trust and mutual respect and became a person Lucas now knew who broke his mold of what academics and scientists do and say.

During conversation, Victor and I shared information about personal subjects and our own experiences. These were opportunities for me to discuss the poor treatment I had received as an academic from climate deniers, which Victor seemed to sincerely apologize for. He also noted instances where he had felt similar exclusion from climate discussions and that he had been previously silenced when bringing up his questions and skeptical beliefs. I apologized that Victor had felt shut out of conversations and used myself as an example of how not all environmentalists or academics participate in silencing or exclusion. I observed that my willingness to listen and sympathize tempered or at least slightly altered this bargainer's preconceived biases toward academics and environmentalists. In our exchange of experiences we had with generalizations, Victor seemed to recognize that his perceptions of climate supporters and academics were not representative, just as he had felt unfairly generalized by the term "denier." He noted that my "interest" in learning more about the skeptical community was genuine and real, which I believe contributed to our free exchange of ideas.

Employing examples is a powerful strategy that can help disrupt associations that bargainers have toward climate scientists, academics, and environmentalists. Instead of letting stereotypes and media drive representations of these groups, we can serve as alternative models of those who will listen, who give trust and understanding, and who value the environment but do so in ways that do not jeopardize the reigning frameworks of the bargainers. Others who engage with bargainers may not be able to draw upon experiences as academics but everyone can draw upon their experiences and use themselves as a counter-example in other ways. For example, environmentalists who are business owners can discuss the negotiation of caring for both the environment and the economy. People of faith can draw connections between their interpretations of the Bible and how they mobilize those passages and verses to love both God and the Earth. Conservatives (and moderates) can share a skepticism of the politicization of climate change and serve as an example of how all people, regardless of political orientation, should care about the potential dangers of climate change.

Strategies to avoid

Because bargainers view themselves as part of the scientific community with competing scientific information, claims that they are unscientific will likely activate defensive frames. We do not want to insult or demean people; everyone has reasons for thinking a certain way. Instead of placing labels on bargainers as unscientific or denouncing their cherry-picked results, we can respond by being open to their perspective and offer our own interpretations. As discussed in the separators' chapter, it is important to embrace a dialogue approach to conversations. Especially considering that the bargainers' framework invites dialogue and deliberation, we should embrace that opportunity for engagement.

Directly addressing the inadequacies and flaws in the bargainers' reasoning may be construed as a form of silencing. But, by joining the revolution and becoming part of the bargainers' community, we can prompt self-reflection and challenge assumptions that bargainers may have about environmental advocates. Scholars have addressed the risks in ignoring climate skeptics as this gives them an upper hand in terms of controlling narratives and maintaining relevance in the public sphere as an undisputed narrative (Ceccarelli 2011: 212). It may also be the case that such dismissals work to foster resentment and generalizations about the closed-mindedness of environmental advocates. While we have no responsibility to (and should not) recognize and acknowledge the legitimacy of unscientific data, interpretations, and authority, we do have the responsibility to respect and acknowledge the values, perspective, and person behind those false interpretations. Victor noted that people's "inability" to have conversations is tantamount to "poison" in a productive society. He continued that there is simply no more work that climate advocates can do to convince skeptics until they start to "listen" to minority opinions. While this exchange could certainly be viewed as further evidence that skeptics simply seek to continue the conversation to delay action, I also think there is a kinder reading. It seemed to me that part of the reason why persuasive efforts have been failing are *because* skeptics knew the other side was not listening. Being a part of the conversation and being heard (even if no one changes their minds) may help temper some bargainers' skepticism. This echoes my point from the separators chapter that we should employ a dialogue approach instead of a monologue approach to climate skeptics (Johannesen 1974). While monologues can lead to aggressive and coercive arguments, dialogues promote understanding, cooperation, and mutual respect (Brockriede 1972).

Another strategy to avoid is to claim that science is perfect, an ultimate authority, or always right. Just as we can challenge the illusory correlations and preconceived notions that bargainers have toward environmental advocates, bargainers can poke similar holes in universal claims about science. Based on conversations detailed above, where climate skeptics doubt consensus and climate models, it seems counterproductive to continually laud the infallibility of science, which provides bargainers more support for their alternative authorities and examples of when climate models have been less than perfectly accurate.

Conclusion

This chapter has examined appropriate ways to engage bargainers based on their rhetorical strategies of constructing a Biblical filter, appealing to scientific standards, and cherry-picking data and experts. These rhetorical features distinguish bargainers from separators and call for different responses. Based on theoretical analysis and case study conversations with bargainers, I propose that conversations with bargainers should follow three core strategies: working within frames, joining the revolution, and employing examples. These strategies acknowledge the values and perspectives of the bargainers while creating rhetorical spaces for critical thinking and reconsidering biases. Similar to my advice for the other two groups, it is imperative to approach bargainers through dialogue. Although separators may be standoffish and defensive instinctively based on their frameworks, my conversations with bargainers were far simpler to begin and continue. I believe that this success was due, at least in part, to my willingness to read bargainer resources, exchange information, and not dismiss their beliefs out of hand. If we can meet bargainers at that same level of respect, our simple openness to engage becomes part of our rhetorical labor.

An important feature of the bargainers is their incorporation of faith into their climate change beliefs. All three groups studied here are religiously-motivated in their climate change attitudes and beliefs, but religion plays different roles to different extents for each group. For example, separators were more likely to evoke and call upon religious justification regularly in our conversations. Bargainers, on the other hand, united their faith with other values such as personal autonomy and neoliberalism, embedding religious language and ideologies in their frame but evoking it more indirectly than directly. Bargainers are more likely to view climate change conversants as misguided dialogue partners who need to be corrected rather than dismissed. We should not mistake this indirect treatment of faith to mean that bargainers are not as faithful as the separators; my conceptualization of these groups is not based on a scale of severity of belief or denial. Indeed, the bargainers so deeply root their beliefs and actions in their faith that it becomes an innate, natural part of their responses. A more accurate way to describe the distinction is that bargainers embrace their faith in a nexus of beliefs that influence their climate change attitudes and actions, while separators more prominently feature overt religious inventional resources. I personally found that bargainers were easier to talk to. They, unlike the separators, seemed genuinely interested in discussion and deliberation. While I would not qualify such conversations as necessary to clarify scientific data, considering the weight of scientific consensus, my bargaining partners clearly saw a need for further discussion and thus were open to having one. These were by far the longest conversations I had, even longer than most harmonizers.

The next chapter explores the final group, the harmonizers, who are not a form of climate skeptic, but may be considered apathetic in terms of their actions. Separators and bargainers reach conclusions about climate change that contradict climate science, but harmonizers represent a religiously-motivated collective that agrees with mainstream climate science but may not feel that they can make a difference (or may not know how to). In examining the harmonizers' rhetorical features, we explore ways that people unite Christianity and environmentalism as complementary ideals and that separating and bargaining are not the only possible responses when it comes to the intersection of Christianity and climate science. This final chapter also serves as a foil to amplify the various

rhetorical features of separators and bargainers and how they are distinct iterations of climate skepticism.

References

Abigail [pseudonym]. (2018) personal communication [digital exchange].

Bloomfield, E.F. and Katula, R.A. (2012) Rhetorical Criticism of the 2008 Presidential
Campaign: Establishing Premises of Agreement in Announcement Speeches. *Communication Research Reports* 29(2): 139–147. DOI: 10.1080/08824096.2012.667775.

Bloomfield, E.F. and Lake, R.A. (2015) Negotiating the End of the World in Climate
Change Rhetoric: Climate Skepticism, Science, and Arguments. In: Meisner, M.S.,
Sriskandarajah, N., and Depoe, S.P. (eds.) *Communication for the Commons: Revisiting Participation and the Environment*, pp. 384–396. Uppsala, Sweden: The International Environmental Communication Association.

Bloomfield, E.F. and Tillery, D. (2019) The Circulation of Climate Change Denial
Online: Rhetorical and Networking Strategies on Facebook. *Environmental Communication* 13(1): 23–34. DOI: 10.1080/17524032.2018.1527378.

Brockriede, W. (1972) Arguers as Lovers. *Philosophy & Rhetoric* 5(1): 1–11.

Brüggemann, M. and Engesser, S. (2017) Beyond False Balance: How Interpretive Journalism Shapes Media Coverage of Climate Change. *Global Environmental Change* 42:
58–67. DOI: 10.1016/j.gloenvcha.2016.11.004.

Burke, K. (1984) *Attitudes toward History*. Berkeley, CA: University of California Press.

Ceccarelli, L. (2011) Manufactured Scientific Controversy: Science, Rhetoric, and Public
Debate. *Rhetoric Public Affairs* 14(2): 195–228.

Climate Central (2018) The 10 Hottest Global Years on Record. Available at: www.
climatecentral.org/gallery/graphics/the-10-hottest-global-years-on-record   (accessed
January 9, 2019).

Dillow, M.R. and Weber, K. (2016) An Experimental Investigation of Social Identification on College Student Organ Donor Decisions. *Communication Research Reports*
33(3): 239–246. DOI: 10.1080/08824096.2016.1186630.

Dixon, G.N. and Clarke, C.E. (2013) Heightening Uncertainty around Certain Science:
Media Coverage, False Balance, and the Autism-Vaccine Controversy. *Science Communication* 35(3): 358–382. DOI: 10.1177/1075547012458290.

Dunlap, R.E. (2013) Climate Change Skepticism and Denial: An Introduction. *American
Behavioral Scientist* 57(6): 691–698. DOI: 10.1177/0002764213477097.

Elliott, K.C. (2014) Anthropocentric Indirect Arguments for Environmental Protection.
Ethics, Policy & Environment 17(3): 243–260. DOI: 10.1080/21550085.2014.955311.

Gabbatiss, J. (2018) Death Valley Just Set the Record for the Hottest Month Ever
Recorded on Earth. *Independent*, August 2. Available at: www.independent.co.uk/
environment/death-valley-hottest-month-ever-heatwave-weather-climate-change-
a8474551.html (accessed January 9, 2019).

Goodwin, J. and Dahlstrom, M.F. (2014) Communication Strategies for Earning Trust in
Climate Change Debates. *Wiley Interdisciplinary Reviews: Climate Change* 5(1):
151–160. DOI: 10.1002/wcc.262.

Gross, A.G. (1994) The Roles of Rhetoric in the Public Understanding of Science. *Public
Understanding of Science* 3(1): 3–23.

Hoffman, A.J. (2011) Talking Past Each Other? Cultural Framing of Skeptical and Convinced Logics in the Climate Change Debate. *Organization & Environment* 24(1):
3–33. DOI: 10.1177/1086026611404336.

Johannesen, R.L. (1974) Attitude of Speaker toward Audience: A Significant Concept for Contemporary Rhetorical Theory and Criticism. *Communication Studies* 25(2): 95–104.

Lakoff, G. (2010) Why it Matters How We Frame the Environment. *Environmental Communication* 4(1): 70–81. DOI: 10.1080/17524030903529749.

Lucas [pseudonym]. (2018) personal communication [digital exchange].

Matthews, P. (2015) Why Are People Skeptical about Climate Change? Some Insights from Blog Comments. *Environmental Communication* 9(2): 153–168. DOI: 10.1080/17524032.2014.999694.

McFadden, B.R. (2016) Examining the Gap between Science and Public Opinion about Genetically Modified Food and Global Warming. *PLOS ONE* 11(11): e0166140. DOI: 10.1371/journal.pone.0166140.

Moran, M.B., Lucas, M., Everhart, K., Morgan, A., and Prickett, E. (2016) What Makes Anti-Vaccine Websites Persuasive? A Content Analysis of Techniques Used by Anti-Vaccine Websites to Engender Anti-Vaccine Sentiment. *Journal of Communication in Healthcare* 9(3): 151–163. DOI: 10.1080/17538068.2016.1235531.

Paliewicz, N.S. (2018) The Country, the City, and the Corporation: Rio Tinto Kennecott and the Materiality of Corporate Rhetoric. *Environmental Communication* 12(6): 744–762. DOI: 10.1080/17524032.2017.1416421.

Pew Research Center (2013) GOP Deeply Divided Over Climate Change. Center for the People and the Press, November 1. Available at: www.people-press.org/2013/11/01/gop-deeply-divided-over-climate-change/ (accessed January 31, 2018).

Ratcliffe, K. (1999) Rhetorical Listening: A Trope for Interpretive Invention and a "Code of Cross-Cultural Conduct." *College Composition and Communication* 51(2): 195–224. DOI: 10.2307/359039.

Steven [pseudonym]. (2018) personal communication [digital exchange].

Taylor, C.A. (1992) Of Audience, Expertise and Authority: The Evolving Creationism Debate. *Quarterly Journal of Speech* 78(3): 277–295. DOI: 10.1080/00335639209383997.

Victor [pseudonym]. (2018) personal communication [digital exchange].

5 Defining the harmonizers
Harmony, collaboration, and acceptance

The Green Bible (2008) is a version of the Bible that has phrases and passages related to the Earth and the environment marked in green. By highlighting many passages, *The Green Bible* shows how much of Christianity's sacred text can be interpreted as relating directly to the environment. For example, 1 Chronicles 16:33, which is highlighted in green, reads, "Then shall the trees of the forest sing for joy before the Lord, for he comes to judge the earth." Other passages, such as the allegory of Heaven growing from a mustard seed and Jesus being mistaken for a gardener after his resurrection (Biema 2008), are also color-coded green. *The Green Bible* includes extra material on advice for turning environmental beliefs into activism based on the highlighted verses. The book was published to encourage Christians to "read the scriptures anew" and to foster the view that "caring for the earth" is both a "calling" and a "lifestyle" for the faithful (*The Green Bible* 2008). *The Green Bible* provides evidence in support of a Biblical basis for merging Christianity with environmentalism. Despite traditional associations between Christianity and anti-environmental attitudes (White Jr. 1967), there are those who challenge this association and advocate for the unity between Christianity and environmentalism. Unlike the separators and the bargainers, this third perspective on the relationship between faith and the environment results in a harmonizing framework that views Christianity and environmentalism as compatible. Harmonizers embrace scientific methodologies and conclusions as tools that confirm their faith, because they view all science, including climate science, as an equal revelation of God that matches His other revelation, the Bible. Thus, harmonizers accept much of mainstream climate science's conclusions, but are driven primarily by their faith to follow them.

All three categories can be distinguished by their rhetorical strategies, orienting metaphors, and guiding frameworks. The harmonizers are markedly different from the separators and bargainers (more so than the differences between the separators and bargainers) because the harmonizers' performance of their religious identity does not prompt a rejection of or skepticism toward climate science. When faced with apparent disagreement, harmonizers turn to scripture and their interpretations of it to see if there is a reason for the misalignment. If God created both the Bible and the mechanisms of science, then they must agree. Harmonizers do not alter their interpretations of climate science (as the bargainers do) but modify their hermeneutics to match scientific knowledge. Harmonizers turn to scripture as an inventional resource for action (as the separators do) but use it to promote harmony and cooperation between faith and environmentalism. Where the separators and bargainers see conflict and revolution, harmonizers see unity. The harmonizers' framework promotes commonalities between Christian values and environmentalism that overcome perceived differences as human error. Walter Lippmann (1982: 307) argued, "A religion which rests upon particular conclusions in astronomy, biology, and history may be fatally injured by the discovery of new truths." While Lippmann's predictions may be true in the eyes of separators, who position climate science as oppositional to their organizing framework, harmonizers are not injured by climate science but are bolstered by its further clarification of God's word.

Much like *The Green Bible*, harmonizers emphasize prominent parts of scripture that directly invoke care for the environment and contribute to an ecological framework. In this vein, harmonizers are not wary of new scientific discoveries or the prominent role of scientists and science in the climate change controversy. Because they accept climate science and believe in climate change, they are not analyzed in this book as a category of climate skepticism. They are included, however, to serve as a refutation of claims that religion necessarily leads to climate skepticism and to propose that belief in climate change does not always equate to activism. Similar to others who are aware of the consequences of climate change, harmonizers may face apathy, confusion, or stigmatization that can cause their belief in climate change to fail to materialize as action. Sometimes referred to as the Creation Care movement or green evangelicals, harmonizers reinterpret Christianity to include environmental values (Bloomfield forthcoming). The separators and bargainers' interpretation of Christianity does not have to be the final, deterministic say for how people of faith engage the environment. Often faced with resentment and minority status within their Christian communities, some harmonizers are silent, inactive, and feel more comfortable taking personal instead of community-level environmental actions.

Before elaborating on strategies for turning harmonizers' existing belief into action in the following chapter, I first analyze patterns and trends among a prominent group that exhibits harmonizer discourse, the Evangelical Environmental Network (EEN), and interviews with members of the Creation Care movement. My analysis of publicly available content and personal conversations provide evidence that the harmonizers are a unique facet of the Christian community that uses the same inventional resources as separators and bargainers but engages distinct hermeneutic practices to reach markedly different conclusions. I (Bloomfield forthcoming) describe the Creation Care movement as an "ecologically-restorative identity" that distinguishes itself in rhetorically and practically meaningful ways from other Christians. Despite polls that confirm that harmonizers make up a small portion of the Christian population (Konisky 2018), the Creation Care movement has gained much media and public attention for the words of its leaders and network building with other environmental groups (Bloomfield forthcoming). The harmonizers were by far the most responsive to my requests for conversation and were the easiest of the categories to contact for interviews and follow-up questions. Showing interest in the environment as a topic for discussion may immediately trigger defensive mechanisms for some Christians, but for others, it can serve as a welcoming beacon to express and celebrate an important aspect of their religious identity.

The primary rhetorical feature of harmonizers is their turn to harmony and cooperation as guiding frames for viewing religion and environmentalism. Harmonizers embrace an ecological worldview where all of God's creation is linked and thus valuable, inverting the hierarchies of the separators and bargainers, who rank social issues and the economy over environmental protection. This linguistic feature can be explained by turning to Burke's ideas of identification, defined as finding the shared substance and commonalities between disparate entities. Harmony and identification thus perform the rhetorical work of embracing differences, overcoming conflict, and uniting disparate ways of knowing. The framework of harmony and identification often leads to harmonizers separating themselves, not from climate scientists, but from Christian climate skeptics and deniers as incorrectly interpreting scripture. For the harmonizers, mainstream climate scientists and Christian climate deniers are not enemies or even fools; they simply have not opened themselves up to the idea of unity between Christianity and environmentalism.

Because of the prominent and overt association of Christianity with climate skepticism, harmonizers also share the rhetorical marker of a minority group sometimes afraid to speak up about their environmental beliefs or take action on them because of potential community backlash. Instead of vocalizing their commitment to both faith and the environment in community or public setting, harmonizers often turn to personal activism where environmentally-friendly behaviors can be kept to the private sphere. Others do not directly engage their environmental attitudes because they feel that individual or community-level actions are not enough to make any meaningful change. Teasing out these rhetorical patterns helps explain why some Christians, who would readily accept their identity as part of the Creation Care movement, would find it difficult to transform those beliefs into meaningful action. In what follows, I describe our exemplar harmonizer, the EEN, outline the harmonizers' guiding framework, and describe three prominent rhetorical features: (1) constructing a framework of harmony, (2) negotiating their Christian identity, and (3) justifying individual-level activism.

The Evangelical Environmental Network

The EEN was an important pioneer in proposing care for creation as an evangelical and Christian mandate. Started in 1993, the EEN (2011: para. 1) "is a ministry that educates, inspires, and mobilizes Christians in their effort to care for God's creation, to be faithful stewards of God's provision, and to advocate for actions and policies that honor God and protect the environment." The current EEN president is Reverend Mitch Hescox, who is a prominent evangelical speaker and a contributor to *Sacred Acts: How Churches are Working to Protect Earth's Climate*. In 2006, many evangelical leaders came together to create the Evangelical Climate Initiative (ECI), which outlined goals and purposes for all evangelicals to sign as agreement with their values. Written and promoted by the EEN, the ECI recognizes the "opportunity and [the] responsibility [of evangelicals] to offer a Biblically based moral witness that can help shape public policy in the most powerful nation on earth, and therefore contribute to the wellbeing of the entire world" (Christians and Climate n.d.: para. 1). Famous religious leaders such as Rick Warren (author of *The Purpose Driven Life*) and Leith Anderson (President of the National Association of Evangelicals) signed this agreement as a pledge to engage with climate protection. The EEN is a pioneer of the Creation Care movement and brought Creation Care to national and scholarly attention (Wilkinson 2012).

The EEN advocates that key phrases in the Bible support the scientific consensus on climate change. For example, they note that Genesis 2:15[1] calls for Adam and Eve to tend to the Garden of Eden, which the EEN interprets as a synecdoche for nature as a whole. Creation Care advocates often extend caring for the environment as complementary to caring for humanity and reinterpret the pro-life campaign to encompass non-human life. The EEN (2011: para. 2) argued that all humans are "called to protect our most vulnerable populations, including unborn children, from the consequences of climate change, pollution, and overconsumption of resources." The EEN offers a series of "pledges" that members and interested non-members can take online, where people can commit to taking individual action on a variety of issues such as reducing the use of plastic straws. Creation Care groups simultaneously call upon the authority of the Bible and the accuracy and predictability of climate science to provide dual support for communicating environmental risk. As a prominent voice in political and public discourse, the EEN offers an alternative perspective on the relationship between faith and the environment and how Christians can join the many other faiths that already incorporate care for creation as part of their tenets (Bloomfield forthcoming).

Guiding terms

The harmonizers are guided by a framework that views Christianity and environmentalism as united. The guiding terms of the harmonizers are transcendence, harmony, and identification, which all point toward the integrative view that harmonizers take toward God's creation. In other words, the war of the separators and the revolution of the bargainers are replaced with notions of collaboration, integration, and cooperation. While separators, and bargainers to a lesser extent, view Christianity and climate science as incompatible, harmonizers argue that the two are compatible and even allies, because an understanding of the natural world God created is an understanding of God as well. Climate science does not oppose or contradict faith; rather, science and religion, are "two complementary ways that God reveals himself to humans" (Levinson 2006: 423). Harmonizers seek to break down perceived barriers between science and religion and adjust their religious interpretations to align with the value they place on the environment. These guiding terms lead harmonizers not to melodrama or tragicomedy, but to the process of transcendence through which the perceived differences between faith and the environment can be reconciled and resolved. Burke (1984: 336) argued that transcendence involves "the adoption of another point of view from which [opposing views] cease to be opposites." Transcendence is thus a symbolic "bridging device" that works to overcome incompatibilities and opposition (Burke 1984: 341). This perspective is markedly different from war and revolution in that it proposes "competing ideologies" can be united through "a promise of larger unities" (Zappen 2009: 280, 281).

Transcendence is related to Burke's (1970) concept of identification, which is the act of seeing shared substance between people and ideas that are, by their nature, distinct and different. Instead of highlighting divisions and points of disagreement, harmonizers take on a frame of transcendence where divisions are overcome by points of agreement, similarity, and common values. Because identification and division are compensatory to one another, attempts at identification necessarily invite a consideration of division. Instead, we can think of the relationship between the two, and the resulting level of consubstantiality, as a matter of emphasis. For example, we can describe separators as emphasizing identification with their faith, which heightens their perceived division from environmentalists. Conversely, harmonizers downplay or minimize the divisions between faith and environmentalism to emphasize their identification as a distinct Christian identity. The emphasis on identification prompts harmonizers to view disagreements between climate science and faith as flaws in their own conceptualization of the two, thereby preserving the harmony between them. Through transcendence, there is no crime committed and thus no reason for victimage or mortification to remove pollution. Burke (1974: 148, emphasis removed) argued that the poetic ideal of transcendence achieves its goals "by attaining a perspective atop all the conflict of attitude." Harmonizers do not embrace conflict, and instead aim to transcend in-fighting and line-drawing by constructing a higher unity.

Constructing a framework of harmony

The harmonizers are so named because of their focus on transcendent rhetoric that unites various ideologies in a unifying whole. In discussing unity and harmony between faith and the environment, musical and sound-related language was frequent in the harmonizers' discourse. Vannini and Waskul (2006: 6) argued that "harmony refers to the combination of two or more simultaneous sounds in a manner that is deemed esthetically pleasing." Where once was disorder and dissonance, harmonizers combine into a melodious compromise and partnership. Environmentalism and Christianity are united in a way that is beneficial to both, stronger together than as separate melodies. The presence of these metaphors can be seen in the language of harmony, unity, integration, and shared origins. In response to Berger's (1977: 56) claim that secularization creates a society that is "religiously unmusical," the harmonizers provide a perspective that advocates harmony and compatibility.

Harmonizers, such as our exemplar harmonizer of the EEN, promote unity and harmony between the values of their faith and the value they see in the environment. Instead of focusing on the tragic frame and seeking out people to punish, the EEN (n.d.-b: para. 14) calls for its members "to affirm the following principles of Biblical faith, and to seek ways of living out these principles in our personal lives, our churches, and society." One of the EEN's (n.d.-a: para. 24) goals is to create "renewed harmony and justice between people" and "between people and the rest of the created world." The parts of God's creation have been unnecessarily separated and should once again be united. The EEN does not conform to White Jr.'s (1967) assertions that Christianity and environmentalism are oppositional to one another and instead chart its own path that transcends those perceived differences. With the overarching goal of transcendence and of avoiding conflict, harmonizers navigate their religious and environmental identities to promote peace, cooperation, and interconnectedness.

In their "Evangelical Declaration on the Care of Creation," the EEN (n.d.-b: para. 17) argued, "the creation which God intended is a symphony of individual creatures in harmonious relationship." Creatures, human and animal, are all instruments in God's orchestra of life that function and live together in a coherent and beautiful unity. The EEN (n.d.-b: para. 19) emphasizes the integration and cooperation between humans, other animals, and nature which are all made up of "the same systems of physical, chemical, and biological interconnections." Hescox (n.d.: para. 2) further emphasized the integration of people and nature by using the body's circulation as a metaphor: "Small problems may result in dangerous pollution as water like blood is interconnected." Appealing to notions of balance, the EEN noted that in "the industrial age, we started burning more and more fossil fuels and thus chang[ed] the delicate balance God created for sustaining life on His creation" (Hescox 2012a: para. 4). This balance is so delicate that even "small percentages upset the created balance and put human life at risk" (Hescox 2012a: para. 5). The EEN thus views all life as interconnected and, when rightly balanced, achieving a state of harmony and mutual existence. But, with the interventions of humanity, that balance is thrown off-kilter and harmony is endangered.

Metaphors of harmony emerged in my personal conversations with harmonizers when they discussed the interconnectedness of ecology and all life. Chloe noted that there is "oneness" and "harmony of nature" that should drive Christians to protect it and referred to the Biblical fable of the "lion lying down with the lamb." For Chloe, the lion and the lamb represented the harmony between various animal life and that those animals (the lion standing in for humans) with more power have control over their actions to be caring and kind instead of violent. Calling environmentalism a "justice issue," Chloe placed her environmental values as part of a larger Christian duty of "caring for the common good." Instead of erecting a hierarchy where care for the environment is a diversion of resources from humans, Chloe saw caring for the environment as part of a larger cause where all life benefits. She also evoked notions of consubstantiality, noting that the beauty of nature comes from its simultaneous "unity and diversity," where the differences between all life only serves to highlight the wonder of God's creation. In this statement, the shared substance of all life being a part of creation is the rhetorical route by which division is transcended by identification.

Another harmonizer, David, also linked all human life through its common substance, noting that "God is making everything and He is calling it good." All life, including human life, "grow[s] the mark of their Creator," infusing all life with sacred importance. Because everything is made from God, David saw everything as connected and as "intertwined," nothing, "God has his thumbprint over everything." The value that harmonizers place on nature is in direct opposition to the hierarchies of the separators and bargainers who ascribe value only to the utility of nature for human benefit. Andrew noted, "creation has a value beyond our use" and by searching for "intimacy with ultimate reality," we may get a better sense of God and his teachings. For some harmonizers, the interconnectedness of life was linked to responsibility. David noted that human and non-human life is "more connected than we'd like to admit," because if we admit it, we are called to do something about it. Terms such as connected, unity, and responsibility were used together to highlight the proposed and preferred actions that logically follow from taking an ecological, transcendent view of life.

In addition to the harmony of all life, harmonizers also transcended perceived divides between (climate) science and religion. For example, the EEN (n.d.-c: para. 17) argued that "intellectual excellence [pursuit of science] and Christian conviction could exist in harmony." In a blog post about the 2017 March for Science, the EEN argued that "defending science" is an integral part of their mission because in order to be good Christians, "we need to know the facts. Science provides those facts" (Goebel 2017: para. 5). Andrew noted that he was primarily concerned with "facts about religion," so saw science as a way to get at those important, divine facts. He continued, "if religion and science contradicts [*sic*], we begin to question how we approach that part of our tradition, not the other way around." Unlike bargainers, who start with the Bible and modify climate science interpretations to match their religious beliefs, the harmonizers use information from climate science to guide their interpretations of the Bible. For example, Anthony noted that scientists "have a grasp on climate change," so Christians are called to recognize and respond to that information. He further said, "we are responsible" for how we react to the information we have already received from climate scientists, where a failure to respond adequately is a personal sin. Hannah agreed with this sentiment in arguing that "not being aware" of the evidence for climate change is a purposeful avoidance of the facts and felt that people "bury [their] head[s] in the sand" to justify being "lazy." For Hannah, avoiding the truth of climate science is "where the sin comes in."

The EEN believes that Christians should embrace environmentalism and should not allow the "fear of science" to detract from environmental engagement

(Hescox 2012a: para. 11). Hescox (2012a: para. 1) argued that groups such as the Cornwall Alliance (CA), our exemplar separator, are "misusing science" which "turn[s] people away from the good news in Jesus." These groups are "blinded ... on the science" and cannot understand "real science" through their "contrarian worldview" (Hescox 2012a: para. 7). Also picking up on the separators' turn to war, Hescox distinguishes separators' discourse from that of the EEN based on separators' willingness to counter, attack, and dismiss the potential for compromise. The EEN (n.d.-c: para. 1) described its actions as "explor[ing] the biblical basis for Christian engagement [and] the science of a changing environment." The EEN (n.d.-c: para. 5) argued that is important to "understand the science" in order to guide appropriate actions toward the environment. The EEN places authority with scientific experts and draws upon their "world-wide consensus" that includes "every national academy of science" to support its environmental decisions (Hescox 2012a: para. 7). Referencing a study by Doran and Zimmerman (2009), which bargainers I spoke to readily dismissed, Hescox (2012a: para. 6) noted that 97 percent of scientists argue that global warming is anthropogenic, and the cause is "not natural cycles, but the burning of fossil fuels." The EEN trusts scientific conclusions and uses them as the roots for their environmental activism. David put his faith in climate scientists to inform him on the environment. He noted that based on the measurements of climate scientists, "there is no question" about the reality and dangers of climate change, so "to be blind to that takes a conscious decision" to ignore the science.

The EEN argues that science has reawakened a concern for the environment. Hescox (2012a: para. 2) wrote, "mounting studies and evidence" provide support that "changes to our global system are happening faster than we could have imagined a decade ago." Scientists and environmentalists describe increased risk and urgency that invite a Christian, moral response. For example, the EEN (n.d.-e: para. 4) called attention to harmful consequences of human action, such as "air and water pollution, species extinction," among others, as evidence of the need for change. Hescox (2012a: para. 5) estimated that over 300,000 people die annually as a result of climate change, which is tantamount to the "degradations of creation" (EEN n.d.-b: para. 5). The EEN (n.d.-a: para. 10) further argued that in the pursuit of progress, people have "forgotten our responsibility to care for [the Earth]." The search for advancement and improvement led to extreme environmental changes and an "unprecedented rapid temperature rise" that is ultimately due to the actions and practices of humanity (Hescox 2012a: para. 4).

For harmonizers, climate science reminds Christians of the interconnectedness of life and motivates them "to address some of the greatest challenges facing God's people and the world today" (EEN n.d.-e: para. 3). Many of life's current issues, "climate change, abortion, God's creatures, land conservation, water pollution, light pollution, mercury and the unborn, GMOs, and more" are all connected "with a common thread" to care for the environment and others (EEN n.d.-a: para. 4). The EEN (n.d.-b: para. 30) thus calls for people to "sustain and heal the damaged fabric of the creation which God has entrusted to us." The

EEN hopes to return creation to its original form and reclaim a more balanced, mutually beneficial relationship. The EEN reprimands other groups whose advocacy is based on "denying the physical laws ordained by Our Creator" and instead advocates "trusting in Christ … to develop clean energy, sustainable food production, and care for all creation" (Hescox 2012a: paras. 10, 9). The EEN argues that they are working toward the environmental goals set forward by groups such as the CA and the Acton Institute (AI) while maintaining alignment with mainstream science.

In addition to finding harmony between all life and between religion and climate science, harmonizers' discourse also contains evidence of transcendence in terms of solving conflict and creating compromises. For example, the EEN (n.d.-b: para. 34) described their stewardship as such: "We invite Christians – individuals, congregations and organizations – to join with us … becoming a covenant people in an ever-widening circle." In this quotation, the EEN extends an invitation to all Christians, including separators and bargainers, to become a member of a covenant aimed at including all Christians in environmentalism. Other Christians are not enemies, but are instead potential partners in the environmental cause. The EEN (n.d.-b: para. 35) encourages its members "to listen and work with" others "with an eagerness both to learn from them and also to share with them our conviction[s]." The EEN notes that solutions to climate change "require all of us coming together to unite" current knowledge about "successes that have already been achieved" (Bodakowski 2012: para. 18). The EEN views the climate change controversy as an opportunity for discussion and mutual learning. In this way, its members embrace the rhetoric of unity and harmony and share the bargainers' turn to discussion and deliberation. One harmonizer I spoke to, Hannah, argued for the importance of dialogue. She noted, we should "not [be] afraid to talk about some of these issues" and encourage other Christians to "lead the charge" in environmental protection. Another harmonizer, Andrew, thought it was important to "meet people where they are" and to "communicat[e] through [personal] networks" and with "humility" to create an ever-growing "constituency" of Christian environmentalists. Although he acknowledged it was a difficult task, Andrew valued communication and collaboration within the Christian community. He noted that "Christianity is big and messy," so people should acknowledge and "respect the fact that these people [climate skeptics] are Christians," although they might reach difficult conclusions about what their faith drives them to think and do.

While harmonizers attempt to transcend differences and gather people together for common purposes, they also recognize the difficulty in gaining conversation partners. In our conversation, David noted, "I think that Christians struggle with this topic [climate change]," so it is "difficult" to enact real change in the community. The EEN similarly lamented that "so many in the evangelical church openly resist" discussion about today's social issues (Hescox 2012b: para. 6). The EEN (n.d.-c: para. 6) argued that these Christians have incorrectly interpreted the relationship between their faith and the environment, noting, "It is actually unbiblical to set one against the other." Where people perceive "irresolvable conflict," there are always "peaceful resolutions" (EEN n.d.-c: para. 11).

Hescox (2012b: para. 6) described the "culture wars" between religious environmentalists and religious climate deniers through a metaphor of old and new wine. The "new wine [evangelical action toward climate change] presses us to reconsider our models, lifestyles, and understanding of the good news in Jesus" (Hescox 2012b: para. 6). The new wine reflects unity in overcoming "the great moral challenge of our generation" (Hescox 2012b: para. 6). Hescox (2012b: para. 12) warns not to "place our trust in 'old wine,'" because it quickly sours, just like the souring effects of conflict and disagreement. Instead, people must be filled with optimism and "allow faith and hope to be your guide" to a new climate future (Hescox 2012b: para. 11). The EEN encourages people to let go of "the past [that] is completely blind to a new future" promised by the new wine (Hescox 2012b: para. 7). To remain in an era of conflict, war, and disagreement is to forestall more sustainable environmental futures guided by faith. In the sense that transcendence provides an alternative way of seeing, harmonizers wished to open the eyes of other Christians to see new configurations of their faith and the environment not tainted by politics.

Hescox (2014: para. 13) evoked transcendent features of rising above and seeking alternative paths when he called evangelicals to "move beyond our past and rise to a better future; Jesus did." Jesus literally rose above material and physical issues, and the EEN can copy those actions figuratively through transcendent language and ideologies. The conflict between environmentalism and Christianity are in the past; the EEN is transcendent, looking upwards, and to the future. The EEN frequently mentions its growing members and the extent of its influence. In a statement made by the EEN in support of the Environmental Protection Agency (EPA), Hescox (n.d.: para. 11) noted that "over 120,000 pro-life Christian supportive comments from 14 states" urged the EPA to continue protecting the nation from water pollution. The EEN has experienced tremendous growth with "over 600,000 people [engaged] with EEN" since 2011 (Smith 2015: para. 3). Their ministry expands beyond just evangelicals and involves all Christians and even non-Christians in advocacy work. The EEN reaches across denominations and unites all Christians under a shared environmental commitment. David seemed to echo the EEN's optimism when he noted, "I think the tide is starting to turn," albeit "very slow[ly]." Even though the tide may be coming in slowly, David noted that the small shift, "gives me hope" that "by God's grace, we are starting to see a generational movement." The framework of unity guides harmonizers to see connections between all life, to see compatibility between climate science and religion, and to seek conversations and coalition-building. While the number of Christians exhibiting environmental beliefs may still not be a majority of the Christian population (Konisky 2018), there is competing evidence that this intersection is still active, growing, and formulating new Christian attitudes on the environment. As Hannah stated, life is "all related, we're all connected, and it's all connected to our faith."

Negotiating their Christian environmental identity

In turning to harmony instead of war or revolution, harmonizers employ a different set of vocabularies for making sense of their faith and the environment. This new configuration can be considered a unique Christian identity (Bloomfield forthcoming), in that it fosters markedly different environmental beliefs and behaviors than other Christians. In negotiating this identity, harmonizers foster identification between Christianity and environmentalism and foster division between themselves and other Christians. This process of consubstantiality, or the simultaneous recognition of shared substance and difference, is a rhetorical term that conceptualizes identity formation and communication as a fluid, dynamic process between individuals and groups (Burke 1970). In the prior section, I emphasized the identification work harmonizers do in linking Biblical interpretations with environmentalist thinking. In this section, I explore how the identification work that harmonizers perform necessarily leads to division between themselves and other Christians who do not share their ecological attitudes. In negotiating their position as religious environmentalists, harmonizers perform a unique iteration of the Christian identity. Although striving for transcendence and cooperation, harmonizers acknowledge that they have not succeeded in uniting Christians around a single narrative. This section provides insight into how the harmonizers locate themselves among dissenting and skeptical voices, while recognizing that their faith is by far the most salient motivator behind their environmental attitudes. Taking both identification and division into consideration reveals the dynamic and sometimes complicated process that harmonizers undergo to negotiate their Christian environmentalist identities.

One point of difference between harmonizers and other Christians is their prioritizing of values such as the economy and social issues. While the CA and the AI focus on the potential economic threats environmentalism poses, the EEN is more skeptical of unfettered economic progress and its effects on the environment. The EEN (n.d.-b: para. 5) noted, "we are thankful for the many benefits provided by our modern, advanced economy," but continued, "our economic progress has been accompanied by considerable environmental degradation." The EEN (n.d.-b) supports the actions of the EPA and its regulations for air, pollution, and water regulations. This perspective differs from the economically driven concerns of the CA and the AI who view the EPA's regulations on businesses as detrimental. The EEN (n.d.-b: para. 9) categorizes the tactics of separators and bargainers as "seeking to weaken or delay the regulations" because the regulations will be "too expensive" for businesses. For the EEN, other Christian groups appear to value businesses and money ahead of people's health and wellbeing. In an article discussing the prevalence of mercury in water due to unregulated pollution, the EEN (n.d.-b: para. 9) invited conversation about "how much our children's health is worth." Hescox (n.d.: para. 6) tied the price of human life to environmental degradation and described "non-enforcement" of the economy as "plac[ing] our children at further risk." Instead of focusing on the success of businesses, the EEN directs its attention toward the environmental

effects on people's health, especially of children and other vulnerable populations. Harmonizers appear to flip value hierarchies as they do not believe the
economy outranks environmental or public health concerns. Hannah lamented
that her fellow Christians were not "experiencing Earth in the way that it was
meant to be experienced" because they kept undervaluing nature.

In terms of social issues, harmonizers also challenge Christianity's seemingly
singular focus on the issue of abortion. Hescox described the EEN's advocacy as
pro-life, because environmentalism protects life "from conception to natural
death" where "anything that affects the quality of life is something that's a pro-
life value" (as quoted in Valentine 2014: para. 10). A few harmonizers I spoke to
believed that there is a call to protect the environment embedded in pro-life terminology. For example, Hannah noted that those that say they "respect life" but
are not environmentalists are contradictory: "You can't pick and choose. [If you
are pro-life,] you got to respect all life." Along similar lines, David asked, "Why
can't we care about the unborn and moral issues ... and care about the gifts God
has given us?" Instead of emphasizing division and viewing caring as a zero-
sum act, Hannah and David united caring for the environment as part of the
larger Christian pro-life mission.

David resented the seemingly solitary focus of Christians on abortion, arguing
that "protect[ing] the unborn" has become the "hill that we have died on." He
characterized the opinions of other Christians as saying, "'until we respect the
sanctity of life, I don't care about the environment,' but this is wrong too as we
can and should care about both." For David, their faith was about more than the
abortion issue, but about larger issues of the importance and sanctity of all life.
He continued that Christians should not rightly consider themselves pro-life
unless that have considered that label in terms of "the whole spectrum of life."
David further noted that we should ask Christians, "do you care about the
orphans, do you care about people in third world countries, do you care about
clean water?" If those answers are no, then David felt that one had fallen victim
to the sin of "fail[ing] to do good when given the opportunity." Julian agreed
that Christians have become distracted by only focusing on abortion, which he
thought was a purposeful tactic by politicians to "manipulate" Christians to vote
for them "just based on the issue of abortion." For Julian, reclaiming his Christian identity was about becoming more than "single-issue voters" by expanding
the topics that inspire Christians to be politically active.

The EEN is also distinguished by its hermeneutics and definition of important
Christian terms such as stewardship and dominion. In response to my (Bloomfield forthcoming) Creation Care survey, Jessica argued, "dominion is so often
misinterpreted as carte blanche permission to do what we want with creation."
Another survey respondent, Robert, noted, "Dominion doesn't mean to destroy;
rightly understood, it implies Genesis 2:15." In responding to a question about
dominion, Savannah noted, "humans are a part of creation and while we may be
the crown of creation, we are closer to the rest of creation than to God." Instead
of the CA's hierarchy where humans are the dominators of the environment,
Creation Care language describes a more protective, guardian role for humanity.

The harmonizers understand dominion and Creation Care as an integral part of their faith.

Riley argued that everyone "will be held to account by God for how we manage his creation." Harmonizers believe that Christians should take action to protect the environment because the science clearly outlines the threat to God's Earth. Zoe echoed this idea of responsibility noting, "If I put someone in charge of something, I would expect them to do the same; thus, God expects us to care [for] the earth wisely." The poor state of the Earth is thus a wake-up call from which harmonizers justify their actions and concerns. Nathan described the relationship between dominion and stewardship as "He calls us to take dominion of the garden, but more as a gardener would, not as a warrior. We are the stewards of the earth, not its conquerors." In addition to the gardener metaphor, Katherine used a carpenter metaphor: "One is not a 'master' carpenter by knocking down house, but by building one that withstands." Caroline argued that God had given the Earth to humanity, so believed that "We are to care for the earth as a gift and keep it healthy for future generations." The environment should not be treated with contempt, ownership, and exploitation, but rather with a caring, tending form of stewardship. Lisa seemed incredulous that Christians could think any differently: "I'm not sure how the verse [Genesis 1:26] can be interpreted that humans can do whatever they want to the Earth, including destroying it." These interpretations of stewardship are disruptions of the separators and bargainers' hierarchies because the harmonizers challenge the notion that humans have the right to exploit and dominate the environment.

The differences between how harmonizers interpret the Bible and how other Christians do was quite known to them. When I asked about the behaviors of climate skeptical Christians, most harmonizers responded with sighs and disappointment, but also with recognition that their pro-environmental perspective is not widely adopted in the community. The twin themes of economics and politics emerged in harmonizer responses about how they differ from other Christians. For example, Anthony argued that people vote to protect their "pocketbooks," which is an understandable standard for decision-making. Chloe agreed, acknowledging, "nobody wants to change, especially when there's money [involved]." She elaborated that sometimes making the best environmental decisions might not be the economically best thing to do, but she prefers to make those choices to help her "sleep at night." Chloe implies that others might think short-term about economic benefits but might not consider the long-term ethical implications of their actions. Harmonizers thus frame economic and ethical considerations as separate and sometimes competing options (we must optimize one at the expense of the other), while bargainers conceptualize the economic option as the most ethical one.

In terms of value hierarchies, harmonizers are more willing to make economic sacrifices due to their environmental values, while bargainers reify economics as the primary standard for reasoning. Represented by the Toulmin model, the differences between the grounds support contradictory conclusions:

Harmonizer model

Claim: We should make environmentally-beneficial choices even if it costs more.
Grounds: Because the environment is more important than economic inconveniences.
Warrant: We should make decisions that align with our value hierarchies.

Bargainer model

Claim: We should not make environmentally-beneficial choices if they are not economically advantageous.
Grounds: Because economics is more important than the environment.
Warrant: We should make decisions that align with our value hierarchies.

The reasoning behind both arguments is thus the same; the same warrant moves us from the grounds to the claim. But, the differences between the value hierarchies lead to dissimilar conclusions.

These small differences in hermeneutics and argument structure lead to incredibly different performances of the same faith, causing friction and potential hostility between the groups. Particularly in the decision of incorporating climate science with Christianity, the harmonizers distinguish themselves clearly from the logic of the separators and the bargainers. Andrew described what he saw as the largest difference between religious environmentalists and skeptical Christians: skeptical Christians "liv[e] in their own universes and echo chambers" where they are closed off to considering scientific arguments or engaging with people from different perspectives. I agree with Andrew that I encountered more resistance from separators and bargainers, but they were, for the most part, interested in discussing, looking for more information, and considering my perspective. Andrew makes the important point that without active attempts by environmentalists to reach out to these communities, separators and bargainers might not have those opportunities to engage with other perspectives and may stay locked in networked echo chambers (Bloomfield and Tillery 2019). In other words, I saw little evidence in my conversations that separators or bargainers actively seek out conversations with environmental advocates (i.e., leave their echo chambers) and thus might not have opportunities to consider alternative perspectives. Harmonizers, on the other hand, are by virtue of their identity surrounded by those they perceive to have different opinions and perspective from them on climate change and the environment.

Another way that harmonizers distinguish themselves is to try and avoid political associations. For example, Hescox emphasized transcending politics when he noted, "climate change is not a liberal issue or any issue other than a people issue" (as quoted in Valentine 2014: para. 3). Climate change should transcend politics and avoid labels of liberal and conservative and instead focus on common concern for the environment. When discussing religiously-based climate skepticism with harmonizers, a majority of those I spoke with lamented the role that politics was playing in climate change deliberation. Chloe argued that it is unfortunate that environmental issues "run up against politics" because focusing on politics ignores the rightful role that religion should play.

David noted that politics and party bickering are some of the biggest obstacles to productive conversation, noting that he wished society could "get rid of the right-wing, left-wing goggles we like to look at the world through." In discussing the problems they faced within his family, Cameron described how Christians oftentimes see "their pastors as leaders but also see their conservative representatives as leaders." Cameron thought that the combination of the two was potentially dangerous because politics was rising to the same level of importance and influence as one's faith leaders. Hannah echoed this concern about the influence of politics by mentioning that climate change "has nothing to do with politics, it's more religious-based than anything." Hannah further criticized how being environmentally active has become associated with being "too liberal." Because Hannah identified as conservative, she rejected the association that all environmentalists are liberal and that climate change policies are inextricable from the "liberal agenda." Instead, Hannah noted that climate change is "not a political issue" and to be skeptical is to make a "conscious decision" to not follow the tenets of Christianity. She further described her beliefs on politics and the environment as not "about left or right, [but] what type of action does God call me to?" By transcending political divisions and divisions between science and religion, harmonizers create a unique group identity, while simultaneously distinguishing themselves from other Christians (Bloomfield forthcoming).

Harmonizers thus negotiate their personal identification with the Christian faith while acknowledging that their interpretations and performance of it vary considerably from others who enact a more traditional Biblical hermeneutics. While everyone I spoke to interpreted the Bible in their own way, they all felt that they were reading the true, intended message of God. For example, when I asked Anthony from where his environmentalism emerged, he noted, "we can read from the Bible" that God calls people to "tend to the garden." David said that he believes in environmentalism, "not in spite of the Bible, but because of the Bible." For David, the Bible contained all of the justification needed to act and for him to feel that environmental protection was "an ordainment" from God. He felt that other Christians "have been walking around with blinders on" to the clear and obvious truth of God's word that supports the unification of Christianity and environmentalism.

Justifying individual-level activism

While the CA and the AI advocate anti-regulation and hesitancy, one of the EEN's (n.d.-b: para. 33) explicit goals is to support "public policies which embody the principles of biblical stewardship." On its website, the EEN regularly reports on and praises "bipartisan action" taken by political officials who are willing to reach across the aisle to "work together" on important environmental issues (Goebel 2018: paras. 6, 5). The EEN makes marked attempts to influence politics and encourages others to use their religion and passion for environmental protection. The EEN pursues global and national partnerships, expanding the potential influence of the Christian community. The EEN has a dual focus on engaging within local church communities and supporting national and global initiatives to protect the environment and future generations. This includes "promot[ing] cooperation of best resources, best technology, and best approaches to these problems" through social and political means (Bodakowski 2012: para. 18).

For example, the EEN offers online pledges and reading material in addition to promoting various campaigns about environmental topics. In 2018, the EEN launched a campaign called "the last straw," encouraging people to take a pledge to reduce their consumption of wasteful plastics. For the EEN, "the mountains of plastic trash in God's oceans is an affront to his glory and a failure of our stewardship" (Hescox 2018: para. 9). The pledge includes two private acts people can participate in, which are to "decline straws at restaurants" and to "phase out my use at home" (Hescox 2018: para. 10). Both of these suggestions are personal choices that people can make relatively privately or discretely by changing their own personal habits. This appears to be a sufficient commitment for the EEN as they note, "each journey of repentance begins with a single step" (Hescox 2018: para. 10). In other words, even engaging in simple, private changes can have meaningful environmental outcomes. But, just under these two suggestions, the EEN prompts readers to "consider going deeper," by prompting less private steps people can take such as "asking local restaurants to only provide straws when customers request them" and "asking elected officials and the local and state levels to phase out the use of plastic straws" (Hescox 2018: paras. 9, 10). These two suggestions are more prominent, public, and political statements that ask for individuals to step beyond their private, personal choices to influences the choices and options of others. The use of the term "deeper" implies that those public choices are unnecessary or advanced advocacy options and that the EEN recognizes the difference between personal decision-making and public performances of environmentalism.

While the EEN advocates for public-level activism in many of its descriptive statements, more often than not in my conversations, harmonizer activism seemed limited to the private sphere. Paul Stern (2000: 409) distinguished between "environmental activism," which constitutes public presentations of environmentalism and "private-sphere environmentalism," which constitutes "the purchase, use, and disposal of personal and household products". The difference between public and private activism comes from the level of attention, personal involvement, and public knowledge of such behaviors. Actions that can be taken in private to help the environment are not subject to potential labeling as an identity marker or potential stigma from anti-environmentalist family, friends, and networks. Tam and Chan (2018: 182) referred to this process as the "concern–behavior gap," where people know there is a need to act but either do not act or are reluctant to act (emphasis removed). Tam and Chan (2018: 182) explained the gap through the concept of "free riders," or that environmentalists are reluctant to act because they fear others will perceive their actions as a reason not to act pro-environmentally themselves. While this may be a feature of the environmentalists Tam and Chan (2018) surveyed, it was not present in the discourse of harmonizers. As I will explain in more detail later, the concern–behavior gap for harmonizers seemed primarily motivated by the potential for negative political associations and stigma from other Christians who were skeptical of climate change.

I noticed a clear divide between the behavior promoted and lauded by the EEN (*both* environmental activism and private-sphere environmentalism) and my dialogue partners' preferences for individual actions as the primary way by which they performed their environmentalism. Harmonizers largely discussed climate change in the language of individual sins and errors, so turned to individual solutions to correct them. For example, Hannah talked repeatedly about "personal accountability" for actions and the "individual choices [we make] everyday" to "cut back" as opportunities for environmentalism. When asked how harmonizers responded to their sense of duty of responsibility toward the Earth, most harmonizers referred to personal choices and sacrifices, such as Diego and Alyssa's decision to go vegan. Hannah further explained her notion of poor stewardship as "using more than you need and not being responsible."

The theme of overconsumption was a very prominent environmental argument for Hannah. She started our conversation by equating anti-environmentalism with gluttony and denouncing gluttony as "one of the seven deadly sins." Included in Hannah's use of the term "gluttony" was consumption in all its forms. For her, gluttony means:

> We should not use or consume more than we need. Yes, this applies to the act of eating but so much more. If we use plastic drink bottles to drink a beverage, recycle it. If we have the choice to purchase a product produced or farmed sustainably, choose the one that used the least resources. It is a different interpretation of gluttony but it is still about responsibly consuming.

Hannah argued that people oftentimes only think of gluttony in term of "excessive indulgences," but if you expand the scope of "excessive" to mean anything "more than you need" and "indulgences" to include consumption in any form, then many human actions and choices could be interpreted as gluttonous. Hannah noted that gluttony is particularly sinful because our overconsumption is not only personally indulgent but is also exploitative and "taking advantage" of the gifts of the Earth.

Gluttony was linked to another deadly sin, greed, in the sense that environmental exploitation is oftentimes done in the pursuit of economic benefit. Julian pointed out the hypocrisy in Christians who are "apathetic about the physical world because they know that they have an eternal life in heaven to look forward to, but unfortunately that doesn't stop them from being money-grubbing." He further argued that overconsumption is inherently greedy because "it takes from others." Against the sins of gluttony and greed, Andrew mentioned that "contemplat[ing] our urges is integral to spiritual practices," and "the very thing that makes us human is our ability to restrain ourselves from things we desire to do." Rooted in Biblical passages about sin and temptation, Andrew praised humans for their ability to rationally choose to abstain from urges. For harmonizers, their concerns about overconsumption, greed, and gluttony seems to stem from their ecological view that everything is connected so that taking some resources for their own comfort meant that someone else had to go without.

Many harmonizers I spoke with prioritized and justified private-sphere environmentalism as the primary way they enacted their environmentalist identity by cutting down on their personal consumption and refusing the sins of gluttony and greed in their own lives. Chloe praised "local level" initiatives such as installing solar panels on her house, recycling, and making use of community resources. Additional personal behaviors I heard included visiting farmers' markets, refusing straws and plastic bottles, using reusable bottles and utensils, and cutting down on food that uses high amounts of water such as almonds and chocolate. Hannah thought it was a shame that "people don't make the connection" that their everyday actions consume "fossil fuels." By monitoring individual consumption, Hannah felt that people could have the most impact. David agreed and expressed that these changes are "a lot simpler than what we make it to be," that there "are simple things people overlook," and that it "take[s] five seconds to recycle a tin can." The theme of simplicity resonated with harmonizers who viewed environmental actions as both the right and proper things to do and actions that be easily incorporated into their lifestyle. David noted that environmentalism "begins on an individual basis," indicating that personal beliefs and convictions were where environmental activism came from. He elaborated that "it shouldn't take a bill from Congress to make you care," arguing that activism should be bottom-up and that public-, political-, or national-level initiatives should not be necessary to spark individual environmental behaviors. Hannah also eschewed blaming politicians, industry, and other national players for our environmental woes. She noted that "you [only] have control over your individual actions," so "placing blame on the industry" is useless because we cannot control those agencies. Hannah further elaborated, "we often try to pin [climate change] on others, [but] you have just as much power to make these choices [as they do]." David repeated almost this exact sentiment, noting, "we can only do what we control." Instead of feeling paralyzed by this reduced sphere of influence, however, harmonizers seem empowered to make those decisions and take full advantage of choices that were in their capacity to make.

I had an extended conversation with Anthony about the tension between personal and public activism. He explained that he personally explores "ways to cut down on usage" to avoid overconsumption. When I asked him to elaborate about other ways he enacted his environmentalism in public, Anthony hesitated.

He noted that he knows he is "not making much of a dent" with his individual choices, such as "try[ing] to avoid eating meat," but was hesitant to make public statements. Anthony admitted that the political differences he sees between Christian environmentalists and climate skeptics prevent him from being bolder. While he was happy to share his beliefs with family and close friends, he felt that public advocacy in his church or local community was out of the question. For fear of discrimination and isolation, many members of the harmonizer community were unwilling to make more prominent public actions. David offered that others might similarly avoid talking about the environment "because it's become politicized." Chloe argued she did not mind focusing on private actions because "a lot of work" gets done "anonymous[ly]." For Chloe, public recognition of environmentalism was less important than making quick and meaningful changes.

Of the harmonizers I interviewed, only one, Andrew, was engaged in regular and prominent public actions such as "blogging about the environment and faith" and directly "work[ing] with the faith community to mobilize that constituency." Andrew is a member of the clergy and used his position within the community to advocate for environmentalism. Unlike other harmonizers who were wary of speaking publicly, Andrew saw his role within his community as necessitating community- and societal-level work. Andrew, along with many survey respondents, were active members in official Creation Care groups, while most of the harmonizers I spoke with via online chats and phone conversations were not actively involved with local or national environmental groups. A few harmonizers mentioned the potential for larger environmental activism projects, although they did not actively pursue them. For example, David mentioned that "churches could get involved legislatively" and that priests and pastors should "be able to speak to" environmental issues when prompted by their constituents. He elaborated that church communities could take up projects such as "recycling, cleaning up a community, [or] adopt[ing] a highway" to integrate a church's community and ministry work with environmental values.

Some harmonizers linked their emphasis on personal values to their reluctance to come out as an environmentalist. Nini noted that there is a lot of shame associated with pro-environmental actions within the Christian community. While I will expand on this idea in the next chapter, I did want to highlight that harmonizers seem very much aware that they are breaking expectations in being an environmentalist within their faith community. I will propose later that the experiences of these harmonizers provide evidence that harmonizers may make up more of a percentage of the Christian community than surveys have captured because of their reluctance to share their environmental identity (Bloomfield forthcoming). In my dialogues, it was clear that there is an expectation and stereotype that Christians are conservative and/or skeptical of climate change. Breaking that norm requires rhetorical work on the part of harmonizers, who may not want to risk damage to their community networks by revealing their pro-environmental attitudes. To put a finer point on this insight, the harmonizer community, despite finding agreement and alignment between their faith and environmentalism, may participate in self-silencing to avoid potential backlash from a community that they perceive to be oppositional and hostile to their environmental values.

Conclusion

The harmonizers distinguish themselves from other Christian groups through their hermeneutics and approach to climate science, climate change, and environmentalism. Harmonizers unite environmentalism and Christianity as mutually supportive and complementary. The harmonizers share discursive resources and a framework that emphasizes the unity of climate science and Christianity, forging identification where other Christians turn to division. Guided by a transcendent rhetoric, the EEN and harmonizers I spoke with seek to overcome the perceived incompatibilities traditionally associated with Christian environmentalism. Their transcendent frame accepts climate science's conclusions as accurate and meaningful. Unlike the war and revolution metaphors of the separator and bargainers, the harmonizers are not interested in changing or overthrowing the current order. Because harmonizers view Christianity and environmentalism as completely compatible and aligned, there is no need to defend or remedy their faith, values, and identity.

The harmonizers engage guiding terms of transcendence, identification, and harmony to collapse differences between Christianity and environmentalism, negotiate their own identity as religious environmentalists, and justify private-sphere environmentalism. These characteristics are distinct from the separators and bargainers' rhetoric and constitute a third perspective on the relationship between Christianity and environmentalism. Instead of rejecting new information and promoting division and instead of stretching new information to fit previously held beliefs, harmonizers provide evidence that Christians can embrace climate science and see their faith as a motivation for climate activism. This approach deals with dissenting voices through transcendence and prompts scholarly intervention into contemporary movements that break from the divisive binaries of war and revolution. In claiming to be both scientifically inspired and religiously dedicated, the harmonizers invoke a poetic drama that prompts new consideration of the Bible as an ecologically-restorative text (with or without green highlights).

In the following chapter, I detail strategies for engaging harmonizers in public activism and for transforming personal, private actions into community values. Previous scholars have outlined the potential positive and transformative role that the Creation Care community could have on environmental discourse (e.g., Bloomfield forthcoming; Prelli and Winters 2009). They have argued that Creation Care attitudes have the potential to create rhetorical in-roads in political discourse about the environment due to its strategic location straddling two inventional resources of faith and climate science. This chapter provides insight that the movement may not be catching on or fully greening Christianity in part because of the stigma from other Christians and the value harmonizers place on personal activism. Harmonizers are perhaps the most useful group to have conversations with as they are open to discussing their values and share a confidence and belief in mainstream climate science. But it is still important to engage this community with rhetorical strategies that acknowledge their unique positioning and potential reluctance to becoming public environmental activists. In addition to addressing harmonizers as a foil to separators and bargainers to reveal the various ways that Christianity and environmentalism intersect, I also include the harmonizers in this study of climate skepticism because of their emphasis on private acts instead of public action, meaning that there are potentially opportunities for increased activism from an open audience willing to engage on environmental topics.

Note

1 "The Lord God took the man and put him in the Garden of Eden to work it and take care of it" (New International Version).

References

Alyssa [pseudonym]. (2018) personal communication [digital exchange].

Andrew [pseudonym]. (2018) personal communication [phone call].

Anthony [pseudonym]. (2018) personal communication [phone call].

Berger, P.L. (1977) Secular Theology and the Rejection of the Supernatural: Reflections on Recent Trends. *Theological Studies* 38(1): 39–56.

Biema, D.V. (2008) The Bible Goes Green for the Prius Age. *Time*, September 18. Available at: http://content.time.com/time/magazine/article/0,9171,1842268,00.html (accessed August 27, 2018).

Bloomfield, E.F. (forthcoming) Ecocultural Identity in the Creation Care Movement: Analyzing Contemporary Performance of Religious Environmentalism. In: Milstein, T. and Castro-Sotomayor, J. (eds.) *The Routledge Handbook of Ecocultural Identity*. New York; London: Routledge.

Bloomfield, E.F., & Tillery, D. (2019). The circulation of climate change denial online: Rhetorical and networking strategies on Facebook. *Environmental Communication, 13*(1), 23-34. https://doi.org/10.1080/17524032.2018.1527378

Bodakowski, M. (2012) A Discussion with Reverend Mitchell Hescox. Berkeley Center for Religion, Peace & World Affairs, August 29. Available at: https://berkleycenter. georgetown.edu/interviews/a-discussion-with-reverend-mitchell-hescox (accessed October 3, 2018).

Burke, K. (1970) *The Rhetoric of Religion: Studies in Logology*. Berkeley, CA: University of California Press.

Burke, K. (1974) *The Philosophy of Literary Form: Studies in Symbolic Action*. Berkeley, CA: University of California Press.

Burke, K. (1984) *Attitudes toward History*. Berkeley, CA: University of California Press.

Cameron [pseudonym]. (2018) personal communication [digital exchange].

Caroline [pseudonym]. (2016) survey response.

Chloe [pseudonym]. (2018) personal communication [phone call].

Christians and Climate (n.d.) ECI Statement. Available at: http://christiansandclimate.org/ statement/ (accessed November 19, 2012).

David [pseudonym]. (2018) personal communication [phone call].

Diego [pseudonym]. (2018) personal communication [digital exchange].

Doran, P. T., & Zimmerman, M. K., (2009) Examining the scientific consensus on climate change. *Eos, 90*(3), 22-23.

Evangelical Environmental Network (EEN) (2011) Available at: http://creationcare.org/index.php (accessed December 1, 2018).

Evangelical Environmental Network (EEN) (n.d.-a) Creation Care Champions. Available at: www.creationcare.org/creation_care_champions (accessed October 3, 2018).

Evangelical Environmental Network (EEN) (n.d.-b) Evangelical Declaration on the Care of Creation. Available at: www.creationcare.org/evangelical_declaration_on_the_care_of_creation (accessed October 2, 2018).

Evangelical Environmental Network (EEN) (n.d.-c) Loving the Least of These. Available at: www.creationcare.org/loving_the_least_of_these (accessed October 2, 2018).

Evangelical Environmental Network (EEN) (n.d.-e) Why Creation Care Matters. Available at: www.creationcare.org/why_creation_care_matters (accessed April 2, 2018).

Goebel, T. (2017) EEN is Marching for Science. Evangelical Environmental Network. Available at: www.creationcare.org/een_is_marching_for_science (accessed January 28, 2019).

Goebel, T. (2018) A Wonderful Display of Bipartisan Cooperation. Evangelical Environmental Network. Available at: www.creationcare.org/a_wonderful_display_of_bipartisan_cooperation (accessed January 28, 2019).

Hannah [pseudonym]. (2018) personal communication [phone call].

Hescox, M. (2012a) My Meeting with Brother Cal. The Earth is the Lord's, Patheos, November 27. Available at: www.patheos.com/blogs/foolsconfidence/2012/11/my-meeting-with-brother-cal/ (accessed October 2, 2018).

Hescox, M. (2012b) New Wine. The Earth is the Lord's, Patheos, April 13. Available at: www.patheos.com/blogs/foolsconfidence/2012/04/110/ (accessed October 2, 2018).

Hescox, M. (2014) Caring for God's Creation is Part of a Pro-Life Ethic. The Earth is the Lord's, Patheos, April 15. Available at: www.patheos.com/blogs/foolsconfidence/2014/04/caring-for-gods-creation-is-part-of-a-pro-life-ethic/ (accessed October 2, 2018).

Hescox, M. (2018) The Last Straw. Evangelical Environmental Network. Available at: www.creationcare.org/the_last_straw (accessed October 2, 2018).

Hescox, M. (n.d.) Water Pollution. Evangelical Environmental Network. Available at: www.creationcare.org/water_pollution (accessed October 2, 2018).

Jessica [pseudonym]. (2016) survey response.

Julian [pseudonym]. (2018) personal communication [digital exchange].

Katherine [pseudonym]. (2018) personal communication [digital exchange].

Konisky, D.M. (2018) The Greening of Christianity? A Study of Environmental Attitudes over Time. *Environmental Politics* 27(2): 267–291.

Levinson, M.H. (2006) Science versus Religion: A False Dichotomy? *ETC.: A Review of General Semantics* 63(4): 422–429.

Lippmann, W. (1982) *A Preface to Morals.* New Brunswick, NJ; Abingdon, UK: Routledge.

Lisa [pseudonym]. (2018) personal communication [digital exchange].

Nathan [pseudonym]. (2016) survey response.

New International Version. (n.d.) Available at: www.biblegateway.com/

Nini [pseudonym]. (2016) survey response.

Prelli, L.J. and Winters, T.S. (2009) Rhetorical Features of Green Evangelicalism. *Environmental Communication: A Journal of Nature and Culture* 3(2): 224–243. DOI: 10.1080/17524030902928785.

Riley [pseudonym]. (2016) survery response.

Robert [pseudonym]. (2016) survery response.

Savannah [pseudonym]. (2016) survery response.

Smith, R.J. (2015) Evangelical Climate Leader Preaches Gospel through Climate Action. Blessed Tomorrow, July 12. Available at: http://blessedtomorrow.org/blog/evangelical-climate-leader-preaches-gospel-through-climate-action (accessed October 2, 2018).

Stern, P.C. (2000) New Environmental Theories: Toward a Coherent Theory of Environmentally Significant Behavior. *Journal of Social Issues* 56(3): 407–424.

Tam, K.-P. and Chan, H.-W. (2018) Generalized Trust Narrows the Gap between Environmental Concern and Pro-Environmental Behavior: Multilevel Evidence. *Global Environmental Change* 48: 182–194. DOI: 10.1016/j.gloenvcha.2017.12.001.

The Green Bible (2008) San Francisco, CA: HarperOne.

Valentine, K. (2014) Evangelical Group: Climate Change Is a "Pro-Life" Issue. Think-Progress, May 20. Available at: https://thinkprogress.org/evangelical-group-climate-change-is-a-pro-life-issue-9fdd2677d04b/ (accessed October 2, 2018).

Vannini, P. and Waskul, D. (2006) Symbolic Interaction as Music: The Esthetic Constitution of Meaning, Self, and Society. *Symbolic Interaction* 29(1): 5–18. DOI: 10.1525/si.2006.29.1.5.

White Jr., L. (1967) The Historical Roots of Our Ecological Crisis. *Science* 155(3767): 1203–1207.

Wilkinson, K.K. (2012) *Between God & Green: How Evangelicals Are Cultivating a Middle Ground on Climate Change*. Oxford; New York: Oxford University Press.

Zappen, J.P. (2009) Kenneth Burke on Dialectical-Rhetorical Transcendence. *Philosophy and Rhetoric* 42(3): 279–301. DOI: 10.1353/par.0.0039.

Zoe [pseudonym]. (2016) survery response.

6 Harmonizer strategies

Shifting frames from private to public, communicating urgency, and thinking globally

Harmonizers are so named because they perceive unity and collaboration between Christianity and the environment where separators and bargainers see war and revolution. When engaging separators and bargainers, I focused on definitions, values, and priorities to try and temper the potentially defensive and dismissive responses and keep the conversation going. When engaging harmonizers, they largely already agreed with the truth of climate science and believed that the Bible encourages, if not mandates, environmentalism. Thus, strategies for having conservations with harmonizers focus on highlighting the benefits of active participation and involvement. In other words, discussions with separators and bargainers are largely about changing mental processes, where we exchange information about worldviews and perspectives. Conversely, discussions with harmonizers are largely about promoting behavior change as a complement to existing beliefs. Strategies for engaging harmonizers center on using that intersection of beliefs to spur behavior change. As discussed in the previous chapter, many harmonizers are already making personal changes based on their identities as religious environmentalists. However, not everyone is making such changes nor are they being public about them because of the potential for ridicule or judgment.

The rhetorical features of the harmonizers are using metaphors of harmony, constructing a unique Christian identity, and justifying individual-level activism. With these rhetorical features in mind, I offer three strategies for engagement in order to encourage more prominent and overt public environmental activism. These strategies emerge from previous literature about strategies for engaging people in environmental activism and my conversations with harmonizers about what motivates them to act. While the strategies provided for separators and bargainers were focused on maintaining conversation and prompting shifts in conceptions to see their beliefs as compatible with environmentalism, the harmonizer strategies primarily target the value of public activism, the urgency to participate in it, and the global impacts of climate change. By expanding the scope of the issue beyond a harmonizer's individual situation, the hope is to encourage harmonizers to act more in public spaces, which can help garner social and public recognition of the Christian environmentalist population. Urging harmonizers to be more open about their perspectives and the harmony they see between their faith and environmentalism may serve to disrupt "illusory correlations" (McFadden 2016) others hold that Christianity has a deterministic relationship with anti-environmental attitudes. There is evidence to show that the reality behind the relationship between religious identity and environmentalism is much more complicated (Bloomfield forthcoming) but when only polarized voices are heard or when alternative narratives are given equal space to legitimate scientific information (Dixon and Clarke 2013), it is easy to see how such assumptions are made and why those who challenge these perceived norms may wish to remain quiet.

Similar to previous strategy chapters, one of the primary goals of engaging harmonizers is to listen and maintain opening, welcome dialogue instead of monologue (Johannesen 1974). Although there is less risk that a harmonizer will disengage from the conversation, it is still important to keep in mind common values of conversation and mutual understanding (Ratcliffe 1999). Most people, including harmonizers, will likely be turned off by arguments that seem to come from a place of coercion or aggression (Brockriede 1972). Instead of listing reasons why harmonizers should be more publicly active, it is imperative to listen to the obstacles harmonizers perceive to their engagement and to locate their particular environmental priorities as gateways to further activism. During my conversations with harmonizers, it was clear to me that even in their over-arching agreement that the environment is important to Christianity, harmonizers still traveled different paths to reach those beliefs and emphasized different features of their faith as part of their environmental attitudes. Harmonizers drew from different Biblical passages, personal experiences, and values to support their identity as environmentalists. Thus, individual attention and tailored messages are still integral components to engaging harmonizers in conversation.

In this chapter, I explore the strategies of shifting frames, communicating urgency, and thinking globally as in-roads to turn belief into behavior and to widen the perspective through which harmonizers view religious environmentalism. All three strategies involve widening the scope of what a harmonizer considers a manageable, meaningful behavior and encouraging them to adopt this behavior quickly. The first strategy, shifting frames from private to public, refers to expanding private-level environmental actions into public displays of environmentalism. In addition to encouraging and praising "private-sphere environmentalism," this strategy works to put environmental beliefs and behaviors in alignment with the more public-oriented "environmental activism" (Stern 2000: 409). The second strategy builds off the previous strategy to communicate the urgency of acting publicly. While harmonizers believe climate change is important and is aligned with the values of their faith, they sometimes lack the motivation to act immediately on these beliefs. This strategy incorporates specific ways of framing the problem of climate change to promote a sense of urgency.

The final strategy is to encourage harmonizers to think globally. Instead of focusing on the differences between private- and public-sphere activism (as we do in the first strategy), this third strategy uses a global view to accrue examples about the harms that other countries and people have already befallen because of climate change's consequences. A global frame invokes the ecological values of the harmonizers while also providing concrete examples to spur action. While there is evidence that climate change is already affecting the United States (Parmesan and Yohe 2003; US Global Change Research Program 2009), other countries have suffered and will suffer worse than will likely be observed in the United States in the near future. Emphasizing this imbalance, also referred to as climate justice (de Onís 2012), can activate harmonizers' sense of duty toward all life, especially life that is especially vulnerable due to the effects of climate change. Together, these three strategies build on one another to leverage the beliefs of the harmonizers into more productive and public environmental behavior. As discussed in the previous chapter, very few harmonizers were actively engaged in public-sphere activism, despite acknowledging the dangers of climate change and identifying themselves as religious environmentalists. Although harmonizers are not climate skeptics in the way that separators and bargainers are, they are still important to engage as part of productive climate conversations toward more sustainable futures.

Shifting frames from private to public

Most harmonizers that I spoke with freely discussed the actions they were taking in their personal lives to perform a pious, environmental identity compatible with their faith. While they all drew on different verses and passages to justify their behavior, most of them seemed satisfied to limiting their religious environmentalist identity to the private sphere. Some did express the view that Christians can and should do more than personal actions by sharing those actions with others. For example, Alyssa noted, "I personally am of the view that we [Christians] need to talk more about what we can do to be better stewards of God's creation." In using the verb "talk," Alyssa implies that there should be conversation and engagement with people beyond themselves. This perspective provides an important in-road for shifting the scope of the harmonizers' frame from individual or personal-level activism to public activism. In other words, we can engage the harmonizers' existing regard for personal actions by encouraging them to share the actions they take with others. Thus, their personal actions are multiplied in their effectiveness with each new Christian that adopts them. Alyssa elaborated on her statement by saying that Christians should encourage each other to "adopt instead of having their own children" and to try going "vegan." These were choices that she had made and wanted to share with her networked spaces. If harmonizers are comfortable making personal decisions, we can prompt them to consider sharing those behaviors with trusted friends, family, and networks in their faith community to encourage more people to adopt similarly positive actions for the environment. In this way, the private, personal space where harmonizers feel comfortable operating is expanded to include more components of what might otherwise be considered public space.

Tam and Chan (2018) argued that trust is an important component of environmental activism, in that people who trust their community more are more willing to be and share that they are environmentalists. While Tam and Chan (2018) associate this correlation with a need to reduce "free riders" in the community, I propose that, for harmonizers, increased trust is associated with less fear of rejection from their peers. To follow Tam and Chan's (2018) arguments about the value of trust and to build off of the power of chaining behaviors through personal networks, it is important to ask harmonizers about who they trust and who they would be willing to share their environmentally-friendly behaviors with without fear of reprisal. This strategy can then be passed along, as the originating harmonizer can ask their friend or family member to share the behavior with someone else they trust (who might not be a part of the original harmonizers' group of trust), thereby chaining out the influence that one harmonizer had in their community.

When prompting harmonizers to consider sharing their environmentally-friendly behaviors with their trusted peers, I received a variety of different responses. When I asked Anthony about how he incorporates his environmental attitudes into his church community, he paused for a very long time before responding that he had never considered enacting environmental activism in his church community before and was pausing out of surprise that such a connection had not occurred to him. Despite being actively engaged in a conversation about how his faith had drawn him to environmental behaviors, he appeared to have genuinely not considered taking his environmentalism into his faith community. Some harmonziers seemed open to being more public about their beliefs, seeing it as a positive way to share their environmentalism without going public about their beliefs. More, however, were still hesitant, citing reasons such as the potential for ostracism and criticism from others, even those that they would otherwise trust in different situations. When addressing these harmonizers, I borrowed a tool that helped me engage bargainers: using indirect arguments (Elliott 2014). I spoke with these harmonizers about how I had engaged skeptical Christians in topics of adopting environmental behaviors by emphasizing the benefits of them outside of the scope of the environment, such as through economic, technological, or security benefits. The harmonizers I spoke with seemed to be mostly wary of the responses they might receive from their separator and bargainer peers, so I gave them the same tool I had used to engage Christian skeptics. In my conversation with Alyssa, I asked her to consider other benefits besides environmental ones for adopting or going vegan. She responded that there are a lot of different reasons to adopt and listed a few, including ministry and caring for those less fortunate, especially when adopting from other countries. For veganism, she offered monetary and health reasons as another incentive as plant-based products are often cheaper and healthier than meat. When I prompted her to consider promoting these activities with her networks through these alternative, non-environmental benefits, Alyssa agreed that those reasons may have more leverage with family and friends than through an environmental approach.

Caroline shared with me her frustration in trying to promote environmental behaviors in her peer networks because of what she viewed as hypocrisy. Caroline noted, "We all want a habitual planet for future generations. We know that climate change and natural disasters disproportionately affect the poor," but people still vote for anti-regulation and anti-environmental candidates and forsake those most vulnerable around the world. For Caroline, her environmental and political values are connected inextricably to her beliefs in ministry to the poor and Christians who do not make this connection are viewed as hypocritical. When I asked Caroline about how she approached family members about these topics, she admitted she sometimes does so aggressively or with a dismissive attitude. I shared with the harmonizer the strategy of dialogue and rhetorical listening (Johannesen 1974; Ratcliffe 1999) to approach people with the goal of mutual understanding. She noted that she had long abandoned trying to engage in such conversations but may try again given this new perspective. This conversation highlighted for me how harmonizers that may be called "apathetic" are less apathetic than disillusioned that they can change the minds of other Christians who think differently than they do. Their perceived apathy, then, is not from a lack of trying, but a lack of success and subsequent lack of engagement. For harmonizers, the strategy is thus not only to convince those who have not reached out to their networks to do so, but also to convince those who have tried and failed to try once more. In arguing that members of Creation Care movements should have been able to fully "green" the faith, we may have underestimated the ingrained norms of these communities and failed to equip religious environmentalists with the skills to make meaningful change.

There is evidence that these interpersonal networks can be influential. For example, Gavin shared with me that he changed his opinion on climate change through an interpersonal interaction. In order to not truncate his experience, I have posted his response in its near entirety from our virtual conversation:

I met a girl who studied astrophysics & music and is really intelligent in general (I fully believe she is going to work for NASA one day). Anyway, she was the first fellow Christian I met who didn't subscribe to the conservative ideology that most folks associate with Christianity. Anyway, I don't remember one specific moment, but over a lot of conversations with her, I really decided to start looking at my beliefs and seeing if I could find arguments in the Bible against them – which I did for quite a few of them. I think the main one that changed my mind on climate change was in Genesis, when God made Adam and Eve masters of the earth and all creatures on it. Basically my thought process was, if we're masters of the Earth, shouldn't we be be [sic] using that power to improve the Earth? Like beforehand I believed that pollution ... was a problem in general, but I thought a lot of the more massive stuff like climate change was just another overblown political talking point. Hearing a detailed scientific explanation from another Christian (in an otherwise non-political conversation) made me think more seriously about how this could have an impact on us.

Gavin's story provides evidence that focusing on the networks of harmonizers and how they might chain out their religious environmental beliefs to trusted family and friends may be an effective route to engage more Christians in environmental conversations.

Although the vast majority of harmonizers kept their environmentalism private, a few had begun making public statements about their environmental choices. These harmonizers serve as powerful examples of the kind of evangelization Christians can do for environmental topics. In later conversations, I brought up the actions and behaviors of my previous dialogue partners[1] as exemplars of the success that others have had and as inspiration for others to attempt changes in their communities as well. For example, Ryan had created a blog where he responded to arguments that his family members had made about why Christians should not be environmentalists and why they were skeptical of climate change. When Ryan had little success convincing certain family members in person, he turned to the Internet as a way to express himself more fully with the ability to cite evidence and provide further reading. Scholars have discussed the immense power of digital spaces and its affordances to enable the sharing and understanding of environmental topics (Cagle and Tillery 2017; Elgesem *et al.* 2015; Luzón 2013). Although the Internet can also be a place for disseminating discourses of skepticism and denial regarding climate change (Bloomfield and Tillery 2019; Matthews 2015), it also provides opportunities for others to express their environmental beliefs and create public resources for others to consume. Ryan noted that he stayed away from citing any environmental or mainstream scientific sources as to avoid immediate rejection from those "who are suspicious of the motives" of organizations such as the Environmental Protection Agency (EPA). This rhetorical choice provides evidence that harmonizers are distinctly aware of the obstacles that Christian environmentalists face and the specific conversation-enders they have come up against when engaging skeptics face-to-face.

Ryan had also started sharing environmentally- and economically-friendly behaviors in his workplace. He argued that since encouraging his workplace to print less paper and digitize some important processes, "our efficiency has increased, our mistakes have decreased, and our printer is rarely used." Ryan, without realizing it, made use of neoliberal and economic arguments as inventional resources and indirect arguments (Elliott 2014) to advocate for environmentally-friendly options. For Ryan, the personal actions he was taking felt insignificant. But, in thinking "beyond personal changes" and instead focusing on collective changes, he argued, "we can band together to form and support organizations, and we can encourage existing organizations (e.g., governments, corporations, and non-profits) to be more proactive." In sharing these stories with me, Ryan was inadvertently modeling the shifting of frames from private to public, whereby private-sphere environmentalism widens its scope beyond the individual and begins to embrace social and community-level networks such as their family, friends and peer networks, church community, workplace, and even digital spaces.

The strategy of shifting frames from private to public is thus a subtle strategy that seeks to include more actions, people, and situations within the harmonizers' perception of private, protected, and trusted spaces. Because harmonizers already share environmental beliefs and act in pro-environmental ways in their personal lives, this strategy praises those behaviors and uses them as foundations for future action. If more individuals can receive environmental information from trusted peers, we may begin a grassroots, community-level movement within Christianity for positive environmental change.

Communicating urgency

The strategy of communicating urgency works to transform belief into urgent calls for prominent and public engagement. There are a few ways to communicate urgency such as invoking common sources of authority, loss frames, and melodrama. The first way to communicate urgency calls upon the credibility of climate scientists, religious leaders, and those who span both domains to provide role models to emulate. Unlike separators (who are likely to reject appeals to authority), we can engage harmonizers by turning to authorities who advocate for environmental protection. Depending on the particular harmonizer, we can turn to people such as Pope Francis, Katharine Hayhoe, the Dalai Lama, Richard Cizik, Karen Baker-Fletcher, and other important players in religious environmentalism (*Grist* 2007). In addition to authority figures and organizations, we can also turn to the Bible and important passages as inspiring urgent decision-making. Although separators and bargainers do not respond well to having the Bible quoted at them, due to their rigid ideas of authority and what "counts" as reasonable data, harmonizers are more open to hearing about various environmental interpretations of the Bible as resources to support their environmental identity.

In order to communicate urgency to harmonizers, I referenced Biblical passages that encourage action, involvement, and a duty to act. Most of these verses came from my initial conversations with the harmonizers themselves. In this way, I used verses I already knew resonated with them as motivation to act. For example, a few harmonizers referenced passages in Proverbs about appropriate actions toward neighbors and others. Gavin responded with Proverbs passages to a question I asked about why he identified as an environmentalist. Gavin mentioned Proverbs 3:27–28, which says,

> Do not withhold good from those to whom it is due, when it is in your power to act. Do not say to your neighbor, "Come back tomorrow and I'll give it to you" – when you already have it with you.

This verse was meaningful to Gavin because it represented someone's capability to act as obligations to do so. Playing on the idea of "tomorrow," I asked him if he felt these verses called him to act today and not put off his environmentalism to tomorrow. Gavin agreed that every action counts and the sooner more people get involved the better. I followed up by asking him, "So, what do you plan on doing today?" He paused to think and then said that he could take that trip to the recycling center that he had been putting off. I agreed that was a good idea and commiserated that my current housing did not provide recycling either, so I needed to make a similar drive.

Another way to capitalize on the urgency of climate change is to promote loss frames in terms of what will be left behind, forgotten, or erased from the beauty of God's Earth if immediate action is not taken. By highlighting shared values, such as their regard for life in all forms and the beauty of God's creation, we can acti-vate loss frames for those valued things being lost in the wake of environmental tragedy. For example, David was very concerned with deforestation and the loss of both plant and animal life due to increased land use for farming purposes. He noted that all life "grow[s] beauty, they grow the marks of their Creator," implying that all life is valuable and meaningful because of its relationship to God. He con-tinued, "God has his thumbprint over everything." Leaning on these connections and ecological values is a way to introduce information about threats to life, human and non-human and propose that in losing nature's beauty, God's thumbprint might be lost as well. Evelyn echoed a similar ecological worldview in connecting various forms of life. She argued that anti-environmental attitudes in the Christian community cause "the extinction of wildlife, the destruction of the oceans and forests, and the deaths of thousands of the most vulnerable people. These actions are definitively non-pro-life." While some harmonizers are already making these connections, we can also encourage this line of thinking in others, thereby promot-ing immediate action to be taken for environmental protection akin to Christian activism toward the issue of abortion.

Frequent strategies that have been implemented throughout my conversations have been to focus on the values and priorities of the individual. When discuss-ing the beauty that might be lost due to climate change, Anthony noted, "I don't really care about polar bears," but when we "see the faces" of vulnerable popula-tions and children who are adversely affected by climate change, that is when he feels most obligated to act. Examples of how plant life, forests, and animals were affected by climate change might thus fall on deaf ears for this harmonizer. As he plainly put it: "I'm a people hugger, not a tree hugger." The valuing of human life above non-human animal and plant life, even while valuing all life, also emerged as a primary feature of Creation Care discourse in my survey responses (Bloomfield forthcoming). Thus, it is important to keep in mind that harmonizers are likely most concerned with threats to human life but may themselves express other values in terms of plant and non-human animal life. For example, Frances wrote in a surey response that the story of Noah's ark includes lesson to protect animal life:

> One of the many lessons from that passage [about Noah's ark] is the truth that the animals of this world ARE companions to us, in more ways than we understand, and they are STUCK on this ark with us that we are busily sinking with our sad, violent and plundering pirate ways.

In this passage, Frances recognizes the power humans have over animals and how animals are oftentimes powerless to act for themselves.

Along with individual case studies of loss, it can be productive to make connections between the apocalypse and climate change. This strategy is a difficult one to employ, as a focus on apocalyptic rhetoric or fear appeals can lead to apathy and feelings of fatalism, or that one's situation is hopeless (Foust and O'Shannon Murphy 2009). But, when properly employed and used as a motivating factor, using the apocalypse to communicate a sense of urgency can also lead to empowerment and a sense of duty to act. Gavin explicitly brought up the relationship between Christian beliefs about the End Times and climate change. He noted that they know of some Christians who are anti-environmental and believe that the world will eventually be destroyed in the apocalypse so there is no need to protect it. He argued,

> My problem with this viewpoint is that it basically allows humans to deny that we're seriously hurting the environment, regardless of whether or not a giant flood or whatever is going to occur. But as Proverbs 28:13[2] says, we shouldn't try to hide our mistakes, but own up to them and try to fix them. On a large scale, this applies to pollution and harming the world that God created for us. If we try to pretend that it's not a problem, eventually we're going to pay a very steep price.

For Gavin, the coming apocalypse does not justify doing nothing but instead compels Christians to fix the damages that humans have done to the Earth. As other studies have shown (Bloomfield forthcoming), a belief in the End Times is not necessarily deterministic of environmental apathy. Harmonizers appear to make use of apocalyptic fears as motivation to act and fix things before the end draws near.

Madeline shared similar ideas about the immediate need for action in light of the Second Coming. She noted,

> Even though God will instantly transform nature when Christ returns, that does not mean Christians should just wait for God to do the work when Christ returns. Sanctification means that God gradually transforms believers so they begin acting now as they will act when they are glorified. Even though God will perfect the character of Christians when Jesus returns, they should begin living righteous lives now through God's transforming power.

Referencing Philippians 2:12,[3] Madeline saw her day-to-day actions as opportunities to become closer to "sanctification," which she defined as acting in accordance with her holy nature and preparing herself for entry into the afterlife. In this way, Madeline saw a denial of stewardship as a rejection of God's intentions to sanctify the individual, while the enactment of stewardship moved people closer to that ultimate salvation.

Using the apocalypse as motivation to act, I prompted a few harmonizers to think about if the Second Coming were to happen tomorrow, what would they have wished they had done. One said that they wanted to start an environmental ministry at their church; another said they were planning on putting in a green-house in their backyard; and another said they had not spent enough time out in nature. By prompting them to consider things they would like to do before a potential world-ending event, the hope is that seeds were planted to act on those behaviors. Instead of making a more overt, potentially coercive statement encouraging them to participate in these actions, I wanted to maintain the values of open dialogue and avoid telling them what to do (Johannesen 1974). I took into account their values and their apocalyptic beliefs in the hopes that some made changes or at least began thinking about whether they should be taking more actions in light of their beliefs.

Another way to encourage viewing climate change through a lens of urgency involves taking a page from the separators' use of melodrama. By heightening divisions between environmental Christians and skeptical Christians, harmonizers can be prompted to think beyond their individual actions to embrace societal-level engagement. Schwarze (2006: 246, 256) argued that melodrama can help reject "environmental issues as matters of personal decision-making or action" to instead frame environmental issues as part of broader "socio-political rela-tions." In drawing a line between Christians who are environmental activists and those who are skeptical, melodrama can resist the narrative that environmental issues can be solved solely through "personal habits, private actions, or con-sumption dilemmas" (Schwarze 2006: 248).

When engaging harmonizers, I asked them what they thought of Christians who were skeptical of the environment. Many used comic language and described Christian climate skeptics as misguided, as having interpreting the Bible incorrectly, as having prioritized other Christian values such as abortion over the environment, or as "missing part of their faith," as Hannah noted. To counter this relatively benign framing of Christian climate skeptics, I spoke to harmonizers about the evidence for climate denial machines that operate at the social and political levels to spread climate denial (Dunlap 2013). By represent-ing the power and influence of Christian climate skeptics, who are opposed to the hermeneutics of the harmonizers, as matters of public activism, harmonizers can be urged to match the efforts of Christian climate skeptics. To invoke melo-drama, we may need to draw upon specific examples and details about how climate skepticism funds anti-environmental public discourse (Jacques *et al.* 2008), disseminates skepticism online (Bloomfield and Tillery 2019), and has infiltrated various levels of government and political decision-making (Glenza 2017). These examples can both emphasize the urgency to act and frame Chris-tian climate skeptics as a formidable foe needing a collective, powerful response.

Overall, harmonizers seemed to be well-versed in the variety of pro-environmental actions they could take but might not have had the sense of urgency or have had external motivation to incorporate them as regular, public practices. By providing specific examples of potential losses to the environment and life, highlighting meaningful Bible passages, and referring to discourse (both scientific and religious) that urged immediacy, I attempted to move harmonizers' existing ecological beliefs into prompt, concrete action.

Thinking globally

As an extension of communicating urgency, providing examples may also serve to frame the issue of climate change as more than something that an individual can solve by bringing into consideration global consequences. The primary strategy in encouraging harmonizers to think globally is to highlight notions of climate justice and the interconnected effects and consequences of climate inaction. Castro *et al.* (2016) provide a useful framework of weak and strong activism, whereby weak activism is individual-level activism that represents a minimum level of acceptable engagement and whereby strong activism includes public-sphere activism and behaviors. Strong activism is preferable to weak activism because strong activism carries with it a sense of duty by levying "a social penalization for failing to uphold a certain social order" (Castro *et al.* 2016: 8). In other words, weak activism is still valuable, but only by addressing climate change through strong activism can the global implications of environmental destruction be solved.

Anthony noted that it was learning about a global phenomenon that encouraged him to get more involved in environmental activism. He discussed the "mass exodus" and "hostile" farming conditions plaguing Bangladesh due to changes in the climate. Anthony had focused in on a specific global example of our climate crisis, which he attributed to having a family member living there. His personal environmental actions were thus placed in a much larger frame by which everyday actions were meaningful but might not be enough to impact the global community. Anthony further argued that humans should never think "individually" or "tribally" on any issue, including climate change, because of how connected the world is. In thinking about environmentalism as a choice between "comfort vs. lives of other people," the decision was an easy one for Anthony.

Because harmonizers themselves brought up the global community and how environmental actions should be meaningful on larger scales, I feel this is a pertinent strategy that can be used in discussions with other harmonizers. We can build on feelings of connectedness and global impact of climate change to provide further justification for immediate climate action. Some parts of the United States are already experiencing effects of climate change, but most people may not recognize these changes or attribute them to climate change. Using global examples, however, can be a complement to communicate urgency by providing concrete evidence that these drastic changes are already occurring, especially for the most vulnerable populations across the world. Scholars who have studied the disproportionate effects of climate change's consequences sometimes characterize it as a new wave or extension of colonialism, because the same hierarchies between the first world exploiting the third world remain (de Onís 2012).

Referencing these examples can also promote a sense of climate justice, which can spur harmonizers to act because of the innate inequality that emerges from climate change's effects. Climate justice advocates argue that "marginalized communities are most affected by [climate] phenomena, [but] their voices are frequently excluded from negotiations conducted by those with greater legislative or economic power" (de Onís 2012: 311). Climate justice thus aims to address such inequality by trying to reduce the effects of climate change on those communities and elevate their voices in climate conversations. Some harmonizers had already begun to link their religious environmentalism with themes found in climate justice advocacy. Nathan noted that his commitment to helping the poor drove him to adopt more environmental behaviors: "It is not the rich first world countries that will be hurt by rising sea levels, and increased temperatures, but the third world, poorer countries that depend more directly on the land for food and sustenance." Appealing to climate justice thus works a dual-pronged strategy. First, appealing to climate justice builds on the harmonizers' ecological network by providing examples of people and communities already affected. Second, appealing to climate justice evokes Christian values of ministry to the poor, the value of all life, and charity to those less fortunate. de Onís (2012: 311) argued that climate justice perceives of our climate crisis "both locally and globally," which I argue makes climate justice an appealing in-road for moving between harmonizers' sense of what areas are already affected by climate change and what areas need immediate attention.

For resources on particular global examples to bring up in conversation, there are many government and non-governmental bodies that publish information on global droughts, famines, extreme weather events, migration, and the spread of disease among other consequences attributable to climate change (e.g., Denchak 2016; GlobalChange.gov n.d.; Jackson n.d.). It is also important to think about locations or demographics that may be particularly meaningful to our dialogue partners to simultaneously build a rapport and locate priorities that might motivate harmonizers to act. Preparing a few examples and being able to provide vivid detail or narratives about the situation may also help to improve the harmonizers' connection to the information (Bushell *et al.* 2017), retention of the information (Braddock and Dillard 2016) and decrease the social distance harmonizers feel between themselves and people already affected by climate change (Hart and Nisbet 2012). As opposed to the strategy proposed for separators (to make things personal), harmonizers already think in terms of personal, private decisions. The three strategies here thus do not focus on changing the harmonizers' beliefs, but expanding their frames in three ways (i.e., publicly, temporally, and globally) to encourage immediate, public action. While separators can be addressed more powerfully by thinking about how they themselves are affected by climate change, harmonizers are better engaged by expanding their consideration of the ecological implications of climate inaction.

In this chapter, I included longer, direct quotations from my dialogue partners to highlight the variety of ways that harmonizers negotiate and understand their duty to the compatibility of Christianity and environmentalism. Before outlining a few strategies to avoid when engaging harmonizers, I want to reiterate how open and welcoming harmonizers were when talking about their faith and the role it played in their environmental attitudes. Despite research that supports an antagonistic relationship (or at least a lukewarm, apathetic relationship) between Christianity and the environment (e.g., Konisky 2018), my research here and in other studies (Bloomfield forthcoming) shows there are also positive and productive intersections between the two. What may be clouding or disrupting recognizing these alternative conceptualizations is that religious environmentalists might be operating mostly on a personal and not public level. The overwhelmingly positive response to my request for conversations, the length, depth, and frequency of our conversations, and the warmth in my conversations with harmonizers communicated quite strongly and powerfully to me the rhetorical opportunities present in the intersection of faith and environmental activism.

Strategies to avoid

Because harmonizers ascribe to both religious and climate science authorities, pulling extra information from those sources will likely be received with an open mind. Something to avoid, however, is to inundate harmonizers, or anyone for that matter, with excessive scientific information. Too much data or technical information is likely to contribute more to disengagement than engagement and may come off as patronizing or as that the harmonizer's dialogue partner cares more about approaching the environment through climate science than through religion. To avoid these assumptions and associations, it is best to limit technical information and instead approach harmonizers through examples, narratives, and common authoritative resources. Many scholars, such as Chaim Perelman (1955) and Walter Fisher (1984), argued that people are largely rational animals, but that rationality is not the only way by which humans make decisions. In talking to public or private audiences, instead of technical ones, it is far better to communicate in more accessible and personal discourse fitting to the situation (Goodnight 1987).

It may be tempting to launch into lists of information and statistics confirming climate change to spur people to action, but I caution that such a tactic is not likely to be successful with the separators, bargainers, nor harmonizers. This is, perhaps, one of the hardest learned lessons of this book: the turn to more science to solve opposition to skepticism and apathy can sometimes exacerbate skepticism or deter engagement. If we truly recognize the power of technical information to go only so far and thus commit to abandoning the information deficit model, then we can explore more productive avenues for engaging various communities and identities in environmental activism. There are far more in-roads to productive engagement with the environment and climate change than through technical information and education alone. While we should not abandon either, we should also not let an overreliance on them "deflect" (Burke 1969: 59) other ways of seeing, knowing, and communicating. Our tendency to use technical information as communicative resources would be best served in helping us cultivate examples of the effects that climate change has already had on the world, instead of having them guide the conversation.

To reiterate an important point made in the section on shifting frames, it is important to not undermine the power of individual-level activism. Because harmonizers tend to view faith and sin as private-sphere matters, discounting this perspective may seem rude or condescending. Instead of framing activism as an either/or (between private and public activism), we should discuss public involvement as a worthwhile and valuable complement to private behaviors and how private behaviors can easily become public ones. I thus advise against downplaying the importance of personal habits and behaviors lest harmonizers adopt perceptions of fatalism and futility in facing the climate crisis. In other words, to characterize the value harmonizers place on private-sphere environmentalism as worthless activities may be to imply that there are no meaningful ways to mitigate climate change. There is evidence that increased fatalism reduces agency and promotes a sense of the future as being out of individuals' hands, which can lead to passivity and apathy (e.g., Barker and Bearce 2013; Foust and O'Shannon Murphy 2009; Leiserowitz 2005). The alternative, then, is to approach activism through a both/and instead of an either/or mindset, where both private-sphere activism and public environmentalism have a part to play in environmental advocacy. This both/and approach can be particularly successful if we can expand the scope of harmonizers' private sphere to include their church community and peer networks as an integral part of advocacy and activism.

Conclusion

This chapter focused on motivational strategies for converting the harmonizers' beliefs into public-facing action in hopes of shifting overarching attitudes toward and assumptions about the relationship between Christianity and environmentalism. Harmonizers help us eschew deterministic assumptions that all Christians are anti-environmentalists because they perform rhetorical work to unite their faith with their ecological beliefs. Thus, the harmonizers represent a distinct way of negotiating one's spirituality with the environment that does not abandon organized faith for a "deep green" religion (Taylor 2004), or as Zach worded it: Christians must worship God, they should not "go overboard" and start to "worship Gaia." Harmonizers's discourse negotiate the retention of some of the anthropocentric language of Christianity and the inclusion of productive ecological rhetorics that make room for pro-environmental attitudes as an integral component of discipleship (Bloomfield forthcoming).

To engage harmonizers in dialogue, I propose three strategies that build directly from the rhetorical features of the harmonizers. The strategy of shifting frames from private to public builds on the harmonizers' pre-existing performance of and regard for individual-level activism. The strategy of communicating urgency builds on the harmonizers' sense of how they negotiate their identity within the larger Christian identity to include immediate and purposeful action as a component of stewardship. The strategy of thinking globally evokes harmonizers' ecological orientation and sense of climate justice as features of their harmonizing worldview. While these strategies are separated here to provide details about each of them, they are by no means exclusive strategies. Our approach to all categories discussed in this book are interconnected, dynamic, and fluid. In moving from the theoretical to the practical, we see how attending to the distinguishing rhetorical features of the harmonizers (in addition to the separators and the bargainers) can inform concrete strategies for engagement. As with the separators and the bargainers, opening up a dialogue and listening to our dialogue partner's individual values and perspectives are the most important steps to engagement (Johannesen 1974; Ratcliffe 1999). Harmonizers may be the most willing to engage in discussions about the environment, but they too will be turned off or feel disparaged by discourse that seems to chastise them or pressure them into specific actions and beliefs.

So far, I have addressed the rhetorical features of three types of intersections between Christianity and environmentalism, which I refer to as the separators, bargainers, and harmonizers, and how to engage people who share those rhetorical features in productive dialogue about the environment. The next chapter summarizes the key points I have made about the differences between these groups but also the similarities in how to approach and engage them in conversation. The following chapter also addresses how my analysis of group discourse and interpersonal conversations can provide important lessons for engaging environmental topics outside of the private sphere and in public forums of communication. It is my hope that the information gathered through this book can not only inform how individuals can address and talk to one another, but also how we can apply concepts of rhetorical listening, dialogue, value hierarchies, premises-building, audience tailoring, and other lessons learned from individual-level dialogues to public discussion and deliberation on the matter of climate change and the environment.

Notes

1 Such references were as general as possible so as not to compromise participants' anonymity.
2 "Whoever conceals their sins does not prosper, but the one who confesses and renounces them finds mercy" (New International Version).
3 "Therefore, my dear friends, as you have always obeyed – not only in my presence, but now much more in my absence – continue to work out your salvation with fear and trembling" (New International Version).

References

Alyssa [pseudonym]. (2018) personal communication [digital exchange].
Barker, D.C. and Bearce, D.H. (2013) End-Times Theology, the Shadow of the Future, and Public Resistance to Addressing Global Climate Change. *Political Research Quarterly* 66(2): 267–279. DOI: 10.1177/1065912912442243.

Bloomfield, E.F. (forthcoming) Ecocultural Identity in the Creation Care Movement: Analyzing Contemporary Performance of Religious Environmentalism. In: Milstein, T. and Castro-Sotomayor, J. (eds.) *The Routledge Handbook of Ecocultural Identity*. New York; London: Routledge.

Bloomfield, E.F. and Tillery, D. (2019) The Circulation of Climate Change Denial Online: Rhetorical and Networking Strategies on Facebook. *Environmental Communication* 13(1): 23–34. DOI: 10.1080/17524032.2018.1527378.

Braddock, K. and Dillard, J.P. (2016) Meta-Analytic Evidence for the Persuasive Effect of Narratives on Beliefs, Attitudes, Intentions, and Behaviors. *Communication Monographs* 83(4): 446–467. DOI: 10.1080/03637751.2015.1128555.

Brockriede, W. (1972) Arguers as Lovers. *Philosophy & Rhetoric* 5(1): 1–11.

Burke, K. (1969) *A Grammar of Motives*. Berkeley, CA: University of California Press.

Bushell, S., Buisson, G.S., Workman, M., and Colley, T. (2017) Strategic Narratives in Climate Change: Towards a Unifying Narrative to Address the Action Gap on Climate Change. *Energy Research & Social Science* 28: 39–49. DOI: 10.1016/j.erss.2017.04.001.

Cagle, L.E. and Tillery, D. (2017) Tweeting the Anthropocene: #400ppm as Networked Event. In: Yu, H. and Northcut, K.M. (eds.) *Scientific Communication: Practices, Theories, and Pedagogies*, pp. 131–148. New York; London: Routledge.

Caroline [pseudonym]. (2016) survey response.

Castro, P., ali Uzelgun, M., and Bertoldo, R. (2016) Climate Change Activism between Weak and Strong Environmentalism. In: Howarth, C. and Andreouli, E. (eds.) *The Social Psychology of Everyday Politics*, pp. 146–162. London; New York: Routledge.

de Onís, K.M. (2012) "Looking Both Ways": Metaphor and the Rhetorical Alignment of Intersectional Climate Justice and Reproductive Justice Concerns. *Environmental Communication* 6(3): 308–327. DOI: 10.1080/17524032.2012.690092.

Denchak, M. (2016) Are the Effects of Global Warming Really That Bad? National Resources Defense Council, March 15. Available at: www.nrdc.org/stories/are-effects-global-warming-really-bad (accessed November 4, 2018).

Dixon, G.N. and Clarke, C.E. (2013) Heightening Uncertainty around Certain Science: Media Coverage, False Balance, and the Autism-Vaccine Controversy. *Science Communication* 35(3): 358–382. DOI: 10.1177/1075547012458290.

Dunlap, R.E. (2013) Climate Change Skepticism and Denial: An Introduction. *American Behavioral Scientist* 57(6): 691–698. DOI: 10.1177/0002764213477097.

Elgesem, D., Steskal, L., and Diakopoulos, N. (2015) Structure and Content of the Discourse on Climate Change in the Blogosphere: The Big Picture. *Environmental Communication* 9(2): 169–188. DOI: 10.1080/17524032.2014.983536.

Elliott, K.C. (2014) Anthropocentric Indirect Arguments for Environmental Protection. *Ethics, Policy & Environment* 17(3): 243–260. DOI: 10.1080/21550085.2014.955311.

Evelyn [pseudonym]. (2016) survey response.

Fisher, W.R. (1984) Narration as a Human Communication Paradigm: The Case of Public Moral Argument. *Communications Monographs* 51(1): 1–22.

Foust, C.R. and O'Shannon Murphy, W. (2009) Revealing and Reframing Apocalyptic Tragedy in Global Warming Discourse. *Environmental Communication: A Journal of Nature and Culture* 3(2): 151–167. DOI: 10.1080/17524030902916624.

Frances [pseudonym]. (2016) survey response.

Gavin [pseudonym]. (2018) personal communication [digital exchange].

Glenza, J. (2017) EPA Wipes Its Climate Change Site as Protesters March in Washington. *Guardian*, April 29. Available at: www.theguardian.com/environment/2017/

apr/29/epa-trump-website-climate-change-peoples-climate-march (accessed November 3, 2018).

GlobalChange.gov (n.d.) Natural Resources. Available at: www.globalchange.gov/browse/federal-adaptation-resources/natural-resources (accessed November 4, 2018).

Goodnight, G.T. (1987) Public Discourse. *Critical Studies in Media Communication* 4(4): 428–432.

Grist (2007) 15 Green Religious Leaders. Available at: https://grist.org/article/religious/ (accessed March 16, 2018).

Hart, P.S. and Nisbet, E.C. (2012) Boomerang Effects in Science Communication: How Motivated Reasoning and Identity Cues Amplify Opinion Polarization about Climate Mitigation Policies. *Communication Research* 39(6): 701–723.

Jackson, R. (n.d.) Global Climate Change: Effects. NASA: Climate Change and Global Warming. Available at: https://climate.nasa.gov/effects (accessed November 4, 2018).

Jacques, P.J., Dunlap, R.E., and Freeman, M. (2008) The Organisation of Denial: Conservative Think Tanks and Environmental Scepticism. *Environmental Politics* 17(3): 349–385. DOI: 10.1080/09644010802055576.

Johannesen, R.L. (1974) Attitude of Speaker toward Audience: A Significant Concept for Contemporary Rhetorical Theory and Criticism. *Communication Studies* 25(2): 95–104.

Konisky, D.M. (2018) The Greening of Christianity? A Study of Environmental Attitudes over Time. *Environmental Politics* 27(2): 267–291.

Leiserowitz, A.A. (2005) American Risk Perceptions: Is Climate Change Dangerous? *Risk Analysis* 25(6): 1433–1442. DOI: 10.1111/j.1540-6261.2005.00690.x.

Luzón, M.J. (2013) Public Communication of Science in Blogs: Recontextualizing Scientific Discourse for a Diversified Audience. *Written Communication* 30(4): 428–457. DOI: 10.1177/0741088313493610.

Madeline [pseudonym]. (2018) personal communication [digital exchange].

Matthews, P. (2015) Why Are People Skeptical about Climate Change? Some Insights from Blog Comments. *Environmental Communication* 9(2): 153–168. DOI: 10.1080/17524032.2014.999694.

McFadden, B.R. (2016) Examining the Gap between Science and Public Opinion about Genetically Modified Food and Global Warming. *PLOS ONE* 11(11): e0166140. DOI: 10.1371/journal.pone.0166140.

Nathan [pseudonym]. (2016) survey response.

New International Version. (n.d.) Available at: www.biblegateway.com/

Parmesan, C. and Yohe, G. (2003) A Globally Coherent Fingerprint of Climate Change Impacts across Natural Systems. *Nature* 421(6918): 37–42. DOI: 10.1038/nature01286.

Perelman, C. (1955) How Do We Apply Reason to Values? *The Journal of Philosophy* 52(26): 797–802.

Ratcliffe, K. (1999) Rhetorical Listening: A Trope for Interpretive Invention and a "Code of Cross-Cultural Conduct." *College Composition and Communication* 51(2): 195–224. DOI: 10.2307/359039.

Ryan [pseudonym]. (2018) personal communication [digital exchange].

Schwarze, S. (2006) Environmental Melodrama. *Quarterly Journal of Speech* 92(3): 239–261. DOI: 10.1080/00335630600938609.

Stern, P.C. (2000) New Environmental Theories: Toward a Coherent Theory of Environmentally Significant Behavior. *Journal of Social Issues* 56(3): 407–424.

Tam, K.-P. and Chan, H.-W. (2018) Generalized Trust Narrows the Gap between Environmental Concern and Pro-Environmental Behavior: Multilevel Evidence. *Global Environmental Change* 48: 182–194. DOI: 10.1016/j.gloenvcha.2017.12.001.

Taylor, B. (2004) A Green Future for Religion? *Futures* 36(9): 991–1008. DOI: 10.1016/j.futures.2004.02.011.

US Global Change Research Program (2009) *Global Climate Change Impacts in the United States: A State of Knowledge Report from the U.S. Global Change Research Program.* Cambridge: Cambridge University Press.

Zach [pseudonym]. (2018) personal communication [digital exchange].

Conclusion
What we learn at the intersection of Christianity and climate change

When I started this book, I aimed to disrupt deterministic assumptions that religious attitudes are necessarily oppositional to environmental activism and to argue that the term "climate change skeptics" is multifaceted, reflects a variety of different perspectives and motivations, and offers the potential for intervention and collaboration instead of rejection. Questioning our assumptions about who skeptics are and the relationship between their faith and the environment opens us up to interrogating the true divisions that keep the public from moving forward on issues of environmental advocacy. In framing the climate change controversy as believers vs. skeptics, we are at least partially guilty of engaging the separators' guiding terms of division, highlighting the things that separate us rather than the things we share. Burke (1984: 166) argued that in a world full of tragedy, critics and scholars must themselves unfailingly turn to comedy as a "corrective." He noted that comedy is the preferred perspective "for the handling of human relationships" because of the consciousness raising and reflection that it provides (Burke 1984: 106). The comic frame "considers human life as a book in 'composition'" and "provides the *charitable* attitude toward people that is required for purposes of persuasion and co-operation" without sacrificing our inherent skepticism and "shrewdness" toward humans (Burke 1984: 171, 166, emphasis in original). Thus, we can be both open and charitable toward skeptics while also retaining our own normative standards that climate change is both real and increasingly dangerous.

With the spirit of comedy, forgiving those who have made mistakes and working toward cooperative and productive futures, one of the primary tenets of this book has been to immerse oneself in the beliefs, ideologies, and guiding values of a variety of different intersections in the relationship between Christianity and the environment and between various iterations of skepticism. Burke (1984: 166, emphasis in original) argued, "In deciding *why* people do as they do, we get the cues that place us with relation to them. Hence, a vocabulary of motives is important for the forming of both private and public relationships." In other words, locating motives provides us with the resources with which to connect to and recognize the value of others for the purpose of forming bonds. By giving trust to my dialogue partners and opening myself up for dialogue (Goodwin and Dahlstrom 2014; Johannesen 1974), I tried to locate the person and their values beyond my initial assumptions. I wanted to engage the person behind the beliefs to open up conversation and, at the very least, offer a new perspective on environmentalism even if no immediate attitude or behavior change occurred. I hope that this book's turn to comedy and understanding opens new doors for exploration into collaborations and alliances in environmental activism. Even if those doors eventually close or are not fruitful, the opportunities they represent should not be overlooked, especially in terms of breaking assumptions and rewriting the narrative of climate skepticism.

As a caveat, I do not want to praise comedy at the expense of ignoring its potential restrictions and limitations. As Schwarze (2006) argued and I highlighted in Chapter 6 when discussing harmonizer strategies, there is sometimes a need to highlight and polarize our differences to encourage action. Given a task as difficult as mitigating climate change, perhaps it is a combination of strategies and approaches, tailored to specific audiences, through which we can produce the most meaningful change. If our overarching frame is comic and looks toward observing ourselves as a unified part of humanity, we might be able to locate points within the comedic drama that need a melodramatic or alternative approach to be fully realized. In other words, we may approach all people through a frame of comedy and charity, but may also shift into melodramatic framings, not of ourselves vs. others, but as a collective society toward certain perspectives, actions, and behaviors that endanger all of us. Similar to how our conversations about climate change will unfold differently for each person in each situation, there may be times where how environmentalists and scholars frame situations will need to be flexible as well. Or, in Burke's (1984: 107) words, "Works vary in their range and comprehensiveness. One [person's] character is another['s] mood," so we must be prepared and ready to adjust when necessary.

As a primary contribution to climate change discourse, I propose that we can conceptualize skepticism as caused neither by ignorance nor malicious rejection. Of course, in some cases, it might be caused and/or influenced by these factors. But, if we shift our perspectives and approaches to allow ourselves the opportunity to discover different reasons, we will likely locate more productive inroads for communicating about climate change. In terms of typology, my proposed schema adds another layer of categorization onto existing ways of understanding climate skepticism. Instead of categorizing skeptics along spectrums of their ultimate conclusion (i.e., the strength of their denial), I aimed to explore the starting points and underlying motivators behind climate change attitudes as useful components of categorization. These categories are not intended to be exhaustive or exclusive but are meant to provide a helpful heuristic for locating potential routes to engagement and understanding in the intersection between Christianity and the environment. I address this particular intersection because many people turn to faith when considering their environmental positions (Ross 2013) and Christian leaders have been prominent voices in environmental politics (Wardekker *et al*. 2009). Because of the prominence of faith in US environmental discourse, this book also aims to provide concrete strategies from both existing literature and through this inquiry that can be implemented. In proposing that there are various ways that Christianity and climate change attitudes intersect, I also make a few secondary contributions. First, I dispel the assumptions that a belief in Christianity determines anti-environmental attitudes. While there is evidence from public opinion polls that Christians are unlikely to be environmental activists (e.g., Barker and Bearce 2013; Konisky 2018), I take issue, in this work and in other publications (Bloomfield forthcoming) that these proxy measures are appropriate ways to account for and make sense of the ways that Christianity and environmentalism intersect. Instead of characterizing climate skeptics or Christians in broad strokes, I used a specific, audience segmentation approach that explores new avenues for understanding.

In this conclusion, I summarize my key takeaways in terms of the differences between separators, bargainers, harmonizers, reiterate the value I see in making such distinctions, and outline the common themes that emerged in the strategies I used to connect with, engage, and motivate Christians around the topic of climate change. My summary includes the defining features of each category as assessed through my online conversations and engagement with the category exemplar organizations, the Cornwall Alliance (CA), the Acton Institute (AI), and the Evangelical Environmental Network (EEN). I also want to address the limitations of this work along with what I see as productive avenues for future investigations that will launch more robust testing and implementation of these strategies for the Christian population and how they might apply to other topics and populations as well. I hope that this inquiry into skeptical and hesitant discourse encourages further explorations into how guiding ideologies and belief systems such as religion influence environmental attitudes.

Summary of key insights

Because of my focus on motives and the way in which people viewed the topic of climate change, I drew on the theories of Kenneth Burke and the vast literature on framing, metaphors, and values to guide my analysis. Instead of differentiating between configurations of climate change beliefs and religion on the basis of strength of denial, I endeavored to locate rhetorical patterns and arguments employed at this intersection that represented different orienting screens. Under this analytical framework, I analyzed the online discourse of prominent groups and conducted interviews with individuals recruited through online conversation requests.

Through my research, three patterns of rhetorical features emerged that have the potential to serve as a hermeneutic resource for exploring how Christians negotiate their faith with environmentalism. These structures relied on distinctive metaphors, themes, and discursive resources that constituted their perspective of and attitude toward mainstream scientific conclusions about climate change. Of the three categories, separators, bargainers, and harmonizers, the first two represent climate change skeptics, who largely doubt or refute the need for immediate, widespread action to mitigate climate change. The third category, the harmonizers, represents an "ecocultural identity" (Bloomfield forthcoming) present within Christianity that recognizes the imperative from both faith and mainstream climate science to believe in climate change. Harmonizers are motivated to see Christianity and environmentalism as aligned but may sometimes be wary of environmental activism and the public performance of their ecocultural identity. Harmonizers are not climate change skeptics, but they may appear to be hesitant, cautious, or detached members of the public.

The first argument structure I located relied on the metaphor of war. This metaphor provided skeptics a lens through which to see environmentalism as antithetical and oppositional to their religious values. The metaphor of war closes off and deflects the ability to see compromise and collaboration and instead promotes discourses of hierarchy, competition, and separation. Upon locating the repeated presence of such rhetorical patterns, I named groups and individuals that tend to refute and doubt climate science because of their perceived bifurcations between it and their faith the *separators*. This term is not meant to convey that separators reject all of science, but simply that they see insurmountable gaps between what their faith tells them and what mainstream climate science tells them about the Earth. The separator category is characterized by three rhetorical features. First, separators tend to control definitions, such as what counts as religious environmentalism, stewardship, Christian, and science. In controlling definitions, separators take the upper hand in determining what is allowed and not allowed in the conversation. Subsequently, separators tend to shift the burden of persuasion onto their dialogue partner and hold them to meeting the separators' standards. The third defining rhetorical feature is that separators appeal to their faith as the ultimate decision-making tool, thereby invalidating other methods of environmental reasoning. Together, these rhetorical features make it difficult to enter conversations when those same authorities are not shared.

In order to engage separators, I offered three rhetorical resources: asking questions, accepting premises, and making it personal. The first resource for engagement is asking questions. It is a simple proposition, but it stems from the separators' tendency to feel that they are not being heard. Although media are buzzing with the "false balance" of an ongoing debate about climate change (Dixon and Clarke 2013) and climate deniers may still be prominent voices in both politics (Holden 2018) and social media (Bloomfield and Tillery 2019), climate skeptics perceive that they are being silenced. By opening ourselves up to listening, we may overcome the initial defensiveness of separators that can emerge when broaching environmental topics. The second strategy, accepting premises, stems from the definitions and values that separators hold dear. Instead of trying to change those strongly held beliefs or definitions, we may coax change or at least encourage reflection by working with them. For example, instead of trying to change separators' minds that human life is more important than non-human animal life, I suggest arguing that environmental policies will result in improvement to the quality of human life and protect the lives of vulnerable humans. This strategy does not have the burden of uncoupling separators from their guiding values and instead focuses on finding alignment between their existing values and pro-environmental attitudes and behaviors. The final strategy for engaging separators is to make the issue of climate change personal and local to their values, authorities, and motivations for acting. By learning about what separators value and who they listen to by asking questions, we can propose connections between environmentalism and existing priorities. Elliott (2014: 244) called this approach one of invoking "indirect arguments," whereby additional benefits are highlighted besides environmental benefits.

The second argument structure I located relied on the metaphor of revolution. This metaphor encouraged skeptics to see current interpretations of climate data as inaccurate, exaggerated, and replaceable. The metaphor of revolution did not prompt overt defensiveness as with the metaphor of war, but it still positioned mainstream climate scientists as contrary to the skeptics' interpretations of reality. These skeptics tended to emphasize the need for more research, the folly of pursuing action without more complete knowledge, and what they perceive to be inconsistencies in climate data and exaggerations of the scientific consensus on climate change. Upon locating the repeated presence of such rhetorical patterns, I named groups and individuals that tend to negotiate and bargain with the tenets of climate science to support their faith the *bargainers*. This term is not meant to convey that bargainers reach perfect compromise with climate science, but instead that they borrow certain aspects of mainstream scientific discourse to frame themselves as legitimate, scientific, and informed voices that challenge the conclusions of mainstream climate science.

Stemming from the metaphor of revolution, bargainers' discourse shared three rhetorical features. First, bargainers tended to construct a Biblical filter by which they accepted or rejected climate science based on its adherence to the bargainers' interpretation of the Bible. In setting up this filter, they also determine what evidence from a variety of different domains, including economics, is appropriate to include in environmental decision-making. Different from separators who control what terminology is used, bargainers try to control what interpretations of existing data they accept as reasonable and rational resources. Subsequently, bargainers perform the difficult rhetorical task of simultaneously appealing to scientific standards of deliberation and research while denying that mainstream climate science meets those standards. By appealing to what Taylor (1992: 279) argues is an "outdated" Baconian interpretation science, bargainers misconstrue the consensus over climate change as an unscientific silencing of legitimate minority viewpoints. Their third defining rhetorical feature is that of cherry-picking by which scientific data, experts, and evidence is selectively represented in order to be aligned with the bargainers preferred actions of hesitation and delay (Ceccarelli 2011).

In order to engage bargainers, I offered three rhetorical resources: working within frames, joining the revolution, and employing examples. The first resource for engagement is working within frames. This strategy acknowledges the competing disciplines of expertise, such as religion and economics, which bargainers use as decision-making resources. Bargainers tended to see capitalism and economic progress as part and parcel of religious liberty and human flourishing. Instead of forcing the conversation to shift over to scientific and environmental frames, I tried to work within the economic and religious frames that the bargainers offered. Similar to the idea of accepting premises with separators, we can engage bargainers by tailoring the arguments we use to fit within the frame they most identify with. Hoffman (2011: 3) argued that faith, technology, and national security are "broker" categories that can span the gap between different perspectives on climate change. Although separators will likely reject the sincerity of those who use religious arguments against them, bargainers may be more open to hearing religious perspectives on the topic. Engaging religion may seem inauthentic, so should be employed with caution. I did offer some religious environmental perspectives when engaging bargainers to try and encourage them to locate various ways to interpret their faith and what they perceived environmentalism to be. Others who are religious might gain even more traction by being able to speak authentically and genuinely about their personal faith and find productive overlap in sharing with bargainers the ways in which their faith guides them to be environmentally active.

For most, a safer strategy that is less likely to shut down conversation is to pivot from science and religion into economics and technology because they are more neutral broker categories. These moves are also not without risk. Myers *et al.* (2012) argued that evoking national security frames can result in boomerang effects among climate skeptics because of perceived issues with the authenticity of the reframing attempt. While such attempts at reframing may be subject to increased skepticism within experimental contexts, interpersonal contexts may dampen such boomerang effects by creating a level of personal intimacy between the dialogue partners (Hart and Nisbet 2012). I will note that none of my dialogue partners withdrew from the conversation or stated that they perceived the conversation to be overtly persuasive once an agreement to converse on the topic had been reached. The warning from this previous work is important (Myers *et al.* 2012), however, in that it cautions us to be aware of the potential drawbacks of certain communicative resources and how we might avoid negative consequences when engaged in conversation.

The second strategy for engaging bargainers involves joining the revolution and trying to prompt a reconsideration of the validity of the minority perspective. Instead of trying to deflect the guiding metaphor, as I proposed with the separators, we can actively engage the bargainers' perspective and encourage them to evaluate whether the conditions for revolution have been met or are reasonable in the context of climate change. For example, in one conversation with a bargainer, I tried to compare the vast amounts of evidence there is for climate change to other scientific topics (such as vaccines), for which there are still naysayers and deniers. By using an argument by analogy, I was hoping to disrupt their association, or illusory correlation (McFadden 2016), between the presence of deniers with the imminence of revolution. In positioning oneself as an ally to revolutionaries, we acknowledge that revolutions are important for the advancement of science while also highlighting the rarity of them and how climate denial does not constitute one. I accept the value the bargainers' place in revolutionary thinking and how important revolutions are in moving scientific information forward. In doing so, I try to refute that current skepticism is constitutive of scientific revolution and promote a reconsideration of the scientific evidence in favor of mainstream interpretations.

The final strategy for engaging bargainers is employing specific examples, especially ones within bargainers' guiding frames, to uncouple their preconceived notions about the people they are launching a revolution against. For example, by substituting their image of a religious environmentalist, an academic, or climate scientist, they may begin to see their own monolithic assumptions as similarly flawed as monolithic assumptions about climate skeptics. Using oneself as a counter-example can be a powerful tool, especially if rapport and good will is established through the use of dialogue. In my conversations, using myself as a counter-example within our interpersonal relationship worked to disrupt previously held biases and to make myself vulnerable to critique as a way to build trust (Goodwin and Dahlstrom 2014). In some cases, bargainers may have developed stereotypes because they do not know anyone in those groups. Providing even one example that challenges these assumptions can lead to reflexivity and a reconsideration of these previously held associations.

The third argument structure I located relied on the metaphor of harmony. The reliance on this metaphor distinguishes this group from the separators and bargainers because harmony denotes collaboration, compatibility, and congruence between climate science and Christianity. This group, which I call *harmonizers*, represent a different configuration of faith and environmentalism. Instead of seeing their faith as oppositional to mainstream scientific conclusions about climate change, harmonizers largely agree with and accept the dangers and severity of climate science. It is important to note that even though harmonizers are not deniers, that does not mean that they are all active environmentalists. Many prefer private-level activism and perform their environmental identity in personal ways. If we were to describe harmonizers using the Yale Program on Climate Change Communication's six Americas categories (Leiserowitz *et al.* 2011), we might call them Cautious or Concerned, where there is tacit recognition of the problem of climate change, but little to no action regarding it. Harmonizers are thus like a majority of Americans (57 percent according to Leiserowitz *et al.* 2016) who know about climate change but have not graduated to engagement. The question then becomes, how might we transform concern and awareness into activism?

Harmonizers share three rhetorical features that stem from the metaphor of harmony. Harmonizers actively work to find compromise and compatibility between climate science and their faith and turn to their interpretation of the Bible to justify pro-environmental attitudes (Bloomfield forthcoming). This rhetorical feature is also apparent in the harmonizers' ecological perspective, where all life is connected by the common ancestry of God. Harmonizers negotiate their Christian environmental identity by distinguishing themselves from other Christians who do not share their environmental framing. Although they may share many of the same values and vocabularies, the hermeneutics and scope of the harmonizers' discourse differ, such as the extension of "pro-life" to encompass more issues than solely abortion. The final rhetorical feature of the harmonizers' discourse is the shared focus on individual-level activism. Some harmonizers chose to engage in personal behaviors because they saw faith as an individual activity that must be met with individual sacrifices and behaviors. Others were worried about going public with their environmental views because they feared such beliefs would be met with judgment from their peer networks. For harmonizers, private-sphere environmentalism is a logical way to align their actions with their beliefs and serves as a defensive measure to be both environmentally active and to preserve their interpersonal relationships.

In order to engage harmonizers, I offered three rhetorical resources: shifting frames from private to public, communicating urgency, and thinking globally. The first strategy builds on the harmonizers' individual-level activism by trying to include both public action and community-level activism within the scope of "private" or "personal" spaces. For example, we can encourage harmonizers to view friends, family, and their church community as part of their private network. This strategy also promotes thinking about environmentalism that can be easily accomplished or incorporated in faith communities such as promoting ministries aimed at park clean ups, recycling events, or changes to how church organizations or workplaces function. Similar to the separators' engagement strategy of indirect arguments (Elliott 2014), harmonizers can feel empowered to share their environmental viewpoints when they pitch a pro-environmental event as having economic, community, or moral benefits, lowering the potential for judgment and rejection. The second strategy for engaging harmonizers is to communicate the urgency of climate change by appealing to common authorities, invoking loss frames, and drawing divisions between active and apathetic Christians to motivate action. The final strategy is to expand the scope of climate change to the global level through providing concrete examples to evoke the harmonizers' ecological attitudes and to emphasize the importance of action beyond the individual level. Such an approach does not minimize the power of individual activism but instead encourages harmonizers to think about the need for collective action for meaningful change to be realized.

Across these three groups, there were similarities but also stark differences in how people's interpretations of Christianity drew them to make decisions about the environment. Depending on the frame through which dialogue partners interpreted their faith radically changed their perspective on the issue of climate change, what "counts" as reasonable and appropriate within discussion, and what potential outcomes were on the table. It is perhaps overly repetitive to address this point of the book once more, but I do think one of the more powerful strategies I propose in this book is the focus on dialogue (Johannesen 1974), identification (Burke 1970), and rhetorical listening (Ratcliffe 1999). Johannesen (1974: 95) proposed that the intention or purpose behind communication "shapes" the message being given.

Furthermore, messages contain a "tone," or the "attitude the speaker reveals towards [their] audience" (Johannesen 1974: 95). People thus inadvertently send the message that they view "the audience as equal, inferior, or superior" while communicating (Johannesen 1974: 96). Because we transmit far more than content when engaging in communication, it is imperative to engage the intention and tone of dialogue, which for Johannesen (1974: 96) is characterized by,

> honesty, concern for the welfare of the Other, trust, genuineness, open-mindedness, equality, mutual respect, empathy, directedness, lack of pretense, non-manipulative intent, encouragement of free expression, and acceptance of the Other as a unique individual regardless of differences over belief and behavior.

Engaging dialogue means that persuasive intents are downplayed, while the intent to listen, understand, and share information are lifted. This does not mean, however, that we abandon persuasion altogether, but simply that we adopt the perspective of persuasion as an act of rhetorical listening and identification (Ratcliffe 1999). Ratcliffe (1999: 197, 203) argued that an emphasis on listening moves us beyond "a guilt/blame logic" to create opportunities "for inventing arguments that bring differences together [and] for hearing differences as harmony." Listening and dialogue are thus complementary approaches to engaging in productive conversations because they both provide new ways of understanding that we could not previously have seen.

Burke (1970) argued that the first step to persuasion is identification, where trust, commonalities, and shared substance become the foundational grounds for opening minds, frames, and vocabularies. But, if we only speak, and never listen, how could we ever determine what those common points might be? In thinking about persuasion and identification as "co-constitutive acts" (Bloomfield and Tscholl 2018), we can understand listening and speaking as integral parts of dialogue and mutual understanding. Without a coercive intention or persuasive goal behind my conversations, I was free to explore the various justifications and motivations behind my dialogue partners' skepticism or reluctance without preconceived ideas about what was standing in their way. For example, if I had approached a separator with overt persuasion in mind, I may have entered the conversation already convinced of their ignorance toward climate science or their loyalty to anti-environmental talking heads and would have clouded my ability to consider the broader rhetorical possibilities for why skepticism emerges and is enacted. If I expect the "capacity" and "willingness" to listen from my dialogue partner, I am obligated to provide it myself (Ratcliffe 1999: 204; see also Goodwin and Dahlstrom 2014).

This research aimed to enact communicative strategies of giving trust and rhetorical listening while entering conversations informed by argumentative and rhetorical resources such as framing, metaphors, and argument mapping. My combined approach to discerning rhetorical patterns and engaging people in dialogue provide a starting point for further research into testing and evaluating the utility of these categories and my proposed strategies for engagement. These categories were formed through my inductive look at how Christians negotiate their understanding of their faith with the overwhelming scientific evidence of climate change. While I entered this research with a moral standard in place, that mainstream science about the presence and severity of climate change is correct, I tried to move beyond my preconceived stereotypes about who climate skeptics and deniers are and the "proper" ways that faith and environmentalism should be integrated. Through rhetorical analyses of exemplar groups' publicly available official discourse and conversations with everyday people, I sought to learn more about the intersection of Christianity and climate change and the various motivations behind why this combination is tenuous, productive, antagonistic, and complicated. In addition to embracing a rhetoric of listening and elevating the importance of rhetoric as a means of producing intimacy, I also wish to relate my findings to traditional rhetoric spaces of public deliberation and politics.

Lessons for rhetoric and public communication

Considering that a large portion of this inquiry focuses on fostering interpersonal conversations with people about climate change, I must address the question of what utility these data and conclusions have for public communication and rhetorical theory. One of the primary contributions I see to taking rhetoric to an individual level is the exploration of potential values, motivations, and hermeneutics that are shared across individuals. If we can locate commonalities on an individual level, these same appeals may be applied to public-focused communication in similar target audiences. The turn to patterned rhetorical responses also applies to determining where productive conversations can be had and where they may be impossible. Similar to other scholars of climate change communication, I advocate that some dialogue partners, such as strident separators, need not be engaged or take up our rhetorical attention. For example, Karin Kirk (2018) argues that some people, whom she calls trolls, will never change their minds on climate change, so we should instead focus on those who are more open and willing to engage. Of the nearly 50 people I had conversations with online, I only had 1 negative interaction with a strident separator who undermined my perspectives, used insults, and derailed the conversation with tricks and riddles. This gives me hope that there are far more people that are willing to engage in conversation than "trolls" who are fully closed-minded. While we should keep dialogue and rhetorical listening at the forefront, we should also selectively attend to those who are truly open-minded and offer the most opportunity for productive discussion.

Considering how complicated the topic of climate change is, exploring how individuals make sense of and understand it can also provide new inventional opportunities. For example, of the new information that I found during this inquiry, two points stuck out the most. First, separators and bargainers may agree with many decisions and policies that benefit the environment (i.e., stopping pollution and investing in green technology), but simply balk at the idea of "climate change" or reject what they feel is ideological baggage that comes with the term.

Decoupling the term "climate change" from these policies can thus be a strategy to engage skeptics in policies and behaviors they already value and feel are important. Second, I was surprised by the myriad ways that harmonizers found compatibility between their faith and climate science. This shows us that there is no silver bullet that will convert religious climate skeptics to religious climate activists. But, this study does provide a variety of Bible verses, religious premises, and linkages between faith and environmentalism as potential in-roads to talk with skeptical audiences. One potential fruitful avenue to explore is the connection between pollution and sins such as pride and greed, because of its application of common religious language to environmental degradation.

Another benefit of audience segmentation and targeting particular audiences is that in doing so, we can identify strategies that work across different perspectives and worldviews. For example, Sangalang and I (2018) argued that there are certain story types that are appealing across the political spectrum by analyzing how people who identified as conservative, moderate, and liberal responded to a variety of story types. By segmenting the audience, we were able to determine both what story types were appealing to target audiences and what stories appealed most across audiences. With these benefits in mind, some studies have specifically called for such engagement with individuals and audience segments (e.g., Cozen *et al.* 2018; Goodwin and Dahlstrom 2014; Hine *et al.* 2014). This meso-level approach splits the difference between crafting a single message for everyone and individualized messages that may be time-consuming and limited in effectiveness by locating target groups for engagement. If we adopt Meyer's (2012: 250) conceptualization of rhetoric as "the negotiation of distance between individuals," then one-on-one conversations have much to tell us about the use to symbols to induce cooperation on challenges facing society today, which we can then apply in our group-level and public-level communication.

These strategies are starting points for practitioners of climate communication, climate scholars, and everyday people to rethink how we approach engaging the various intersections of climate change and Christianity. I have theorized and put into practice these strategies solely within the intersection of Christianity and climate change, but I encourage future scholars to test their utility in a broader array of contexts and in public communication. During a few of my conversations with those who responded to my request for interviews with Christians, a few climate deniers professed that they were not Christian but did want to share their thoughts on the environment (and a few wanted to reassure me that religious belief had nothing to do with their skepticism). Based on these interactions, which were productive and extensive, I hypothesize that some of these strategies can be universally applied beyond Christian dialogue partners to engage in lengthy and productive conversations about climate change.

Limitations and future research

In the spirit of fostering useful, versatile, and meaningful knowledge, I will now highlight some of the limitations of this current book and how those gaps should

encourage further research. One limitation is that while the strategies I have proposed have been theorized across various literature including my own work, they
have only been tested in limited capacities. This research seeks to offer potential
avenues for engagement, but they are by no means foolproof, effective in all
scenarios, or without risk. Instead, I argue, given my analysis of existing literature on strategies for engaging climate skeptics and my research in this book,
that these are good starting points for approaching difficult conversations on the
environment. I wanted to use an inductive approach, so I went into conversations
willing to listen and adapted my responses as I went. Thus, I did not enter conversations with persuasion in mind, nor did I measure the effects of the conversation on my dialogue partners in a way that would afford quantitative or
empirical comparison. Such testing is beyond the scope of this book but is an
important counterpart for those scholars interested in drawing statistically significant and group-level effects.

Another limitation of this book is that I may be accused of falling victim to
the same overcategorization that I aimed to avoid. I do not wish to propose that
these categories are exhaustive or exclusive, but to offer them as guidelines and
give a new perspective on understanding the reasons why climate change denial
still pervades public deliberation. As a means of distinguishing between those
who seem to primarily draw from frames of war, revolution, and harmony, I
created the labels separator, bargainer, and harmonizer, respectively. Because
our frames are often deterministic of our attitudes and behaviors, it seems to be a
logical way to distinguish the various conceptualizations of climate change discourse. Other scholars, however, are less convinced of the persuasive power of
frames as a way to address climate change, because of the potential for adverse
reactions (Myers *et al.* 2012). In this book, my conversations sometimes resulted
in rejection and a boomerang effect (Hart and Nisbet 2012), such as in the case
of the strident separator, but more often than not, the non-study environment was
conducive to engaging dialogue partners on the topic of climate change.

It is also important to highlight how my study has focused on pragmatic strategies at the expense of ideological agreement. In other words, while these strategies promote engagement and the shifting of frames, they do not (or at least
have not been measured to) produce adherence to environmentalism and its
values. Indeed, I advocate avoiding such overt labeling for both separators and
bargainers. Such moves are justified in that this book aims to foster dialogue and
engage Christians and skeptics in extended conversations about environmental
topics. But, in focusing on the practical aspects of environmental attitudes and
behaviors, true ecological adherence is in question. I have not, for example,
transformed separators or bargainers into full environmentalists, nor have I
erased the anthropocentric framings often latent even in the discourse of harmonizers (Bloomfield forthcoming). In pursuit of engagement and behavior modifi-
cations, I have, by necessity, downplayed attempts to encourage the full adoption
of an ecologically-restorative attitude. Such work is incredibly important, and I
encourage future scholars to work on strategies for the adoption of such frames.
For this study, I wanted to meet people where they were and promote strategies

that refigured environmental activism so that people might adopt them quickly and willingly.

I believe there is value in incrementalism as a way to reduce boomerang effects and promote a variety of different ideas, behaviors, and attitudes that can be put toward eco-restorative purposes. This move is not without its serious drawbacks, however, as a focus on outcomes glosses over the cultural, classist, racist, and sexist fallout that can occur from non-ecological frames (e.g., de Onís 2018; Endres 2012). Framing environmental issues as tangential or parallel to economic and technological progress undermines the unique considerations that are and should be prominent threads within environmental discourse. To borrow terminology from social movement studies, we might ask if the problem of climate change is dire enough to settle for achieving our instrumental goals (i.e., practical policy and behavior change) or if we should focus primarily on prefigurative goals (i.e., attitude and belief change) (see Juris 2008). Maintaining prefigurative alignment is difficult because compromises to ideological goals are oftentimes necessary to reach compromise and progress on instrumental goals. I do not mean to say that environmental ideologies and values should be sacrificed, but I do offer that maintaining instrumental success at the expense of the prefigurative may be the lesser of two evils when the alternative is no progress at all toward climate mitigation.

A final limitation is also a great strength of this book: its focus on Christianity and its relationship to the environment. Exploring this intersection fulfills a gap in the literature and narrows the complicated topic of climate change to a dynamic and important community within the larger constellation of climate communication. In choosing to select Christians as the focus of my study, I necessarily deflect other groups that would also benefit from having their communication strategies examined. My reasoning for choosing Christians as my primary focus was to engage the potential for audience segmentation (Goodwin and Dahlstrom 2014; Hine *et al.* 2014), where messages are closely tailored and explored in specific contexts. I also chose Christianity because it remains a politically powerful voice in environmental discussions, both for and against environmental activism, with the hope that this book could provide useful information about including faith in our climate conversations.

Despite these limitations, I do believe my book contributes something of importance to environmental communication and rhetorical scholars, practitioners of environmental communication, and the general public. Environmental communication and rhetorical scholars can benefit from the expansive breadth of rhetorical data analyzed and the unification of qualitative interview data with traditional rhetorical analysis. This strategy embodies calls for communication scholars to be more locally-oriented and community-engaged. I have written the book to be theoretically rigorous but also accessible, because I hope practitioners of environmental communication, from politics, non-profits, journalism and media professions, and other institutions can consume and begin to apply these strategies in their messages and outreach. I also hope that these strategies will disseminate into public discourse and be taken up by people in their everyday conservations, both offline and online, with friends, family, peers, and network members that are skeptical, doubting, or reluctant to act on the topic of climate change.

In response to these limitations, I propose four potential areas for future inquiry: (1) applying this study's proposed strategies via experiments or experimental surveys to test their efficacy in controlled environments, (2) applying these strategies to groups or individuals that seem to traverse or complicate my categorization to refine and expand the proposed typology, (3) applying these strategies to climate skeptical identities beyond religion, such as political and regional identities, within climate change communication, and (4) applying these strategies in additional contexts beyond our climate crisis that are met with public skepticism. Other strands of future research may also develop, but these four are proposed as a summary of some of the important questions I feel still need to be addressed by further inquiry.

Conclusion

The conversation about climate change is likely to be an unending one, as humans will constantly need to renegotiate their tenuous and fluctuating relationship with nature. There have been many studies over the years examining different facets and components of the "wicked problem" of climate change (Cagle and Tillery 2015: 147), and we have only scratched the surface in terms of relevant and productive intersections to explore. This book is firmly rooted in the intersection of Christianity and environmentalism as a meaningful connection in contemporary environmental discourse. There are many other intersections that have been and should continue to be explored to provide a fuller picture of the dynamics and features of the climate controversy and how people make sense of their identity and their understanding of the world in a time of crisis. Although Endres *et al.* (2016) warn against solely focusing on the climate crisis, it is hard to imagine environmental communication taking place without operating under the influence and presence of a crisis frame. Instead of avoiding or deflecting climate change, I instead propose that it is important to position our scholarly work meaningfully in relation to that crisis with the knowledge that many other factors may also influence environmental communication.

In my many conversations, I can say that I made true connections with people, had many pleasant exchanges, and opened myself up to considering climate skeptics not as ruthless, ignorant, or dismissive (although it is true that some are), but exploring the implications of climate change science and climate discourse on their functioning in everyday life. When engaging harmonizers, I was surprised to find so many different justifications for the unity of Christianity and environmentalism, which fills me with the optimism the greening of Christianity may still come. To end, I want to turn to the words of Julian, one of the harmonizers, who argued that we should all "be stewards of the Earth, Christian or not." No matter where our beliefs come from, we share a common duty to right our current disruptive role in human–nature relationships. By listening, opening ourselves up to new perspectives, and being willing to engage, we may change the world. At the very least, we may have more conversations, both speaking and listening, with that end in mind.

References

Barker, D.C. and Bearce, D.H. (2013) End-Times Theology, the Shadow of the Future, and Public Resistance to Addressing Global Climate Change. *Political Research Quarterly* 66(2): 267–279. DOI: 10.1177/1065912912442243.

Bloomfield, E.F. (forthcoming) Ecocultural Identity in the Creation Care Movement: Analyzing Contemporary Performance of Religious Environmentalism. In: Milstein, T. and Castro-Sotomayor, J. (eds.) *The Routledge Handbook of Ecocultural Identity.* New York; London: Routledge.

Bloomfield, E.F. and Tillery, D. (2019) The Circulation of Climate Change Denial Online: Rhetorical and Networking Strategies on Facebook. *Environmental Communication* 13(1): 23–34. DOI: 10.1080/17524032.2018.1527378.

Bloomfield, E.F. and Tscholl, G. (2018) Analyzing Warrants and Worldviews in the Rhetoric of Donald Trump and Hillary Clinton: Burke and Argumentation in the 2016 Presidential Election. *Kenneth Burke Journal* 13(2): n.p. Available at: http://kbjournal. org/analyzing_warrants_bloomfield_tscholl (accessed December 1, 2018).

Burke, K. (1970) *The Rhetoric of Religion: Studies in Logology.* Berkeley, CA: University of California Press.

Burke, K. (1984) *Attitudes toward History.* Berkeley, CA: University of California Press.

Cagle, L.E. and Tillery, D. (2015) Climate Change Research across Disciplines: The Value and Uses of Multidisciplinary Research Reviews for Technical Communication. *Technical Communication Quarterly* 24(2): 147–163. DOI: 10.1080/10572252. 2015.1001296.

Ceccarelli, L. (2011) Manufactured Scientific Controversy: Science, Rhetoric, and Public Debate. *Rhetoric Public Affairs* 14(2): 195–228.

Cozen, B., Endres, D., Peterson, T.R., Horton, C., and Barnett, J.T. (2018) Energy Communication: Theory and Praxis towards a Sustainable Energy Future. *Environmental Communication* 12(3): 289–294. DOI: 10.1080/17524032.2017.1398176.

de Onís, C.M. (2018) Energy Colonialism Powers the Ongoing Unnatural Disaster in Puerto Rico. *Frontiers in Communication* 3: n.p. DOI: 10.3389/fcomm.2018.00002.

Dixon, G.N. and Clarke, C.E. (2013) Heightening Uncertainty around Certain Science: Media Coverage, False Balance, and the Autism-Vaccine Controversy. *Science Communication* 35(3): 358–382. DOI: 10.1177/1075547012458290.

Elliott, K.C. (2014) Anthropocentric Indirect Arguments for Environmental Protection. *Ethics, Policy & Environment* 17(3): 243–260. DOI: 10.1080/21550085.2014.955311.

Endres, D. (2012) Sacred Land or National Sacrifice Zone: The Role of Values in the Yucca Mountain Participation Process. *Environmental Communication: A Journal of Nature and Culture* 6(3): 328–345. DOI: 10.1080/17524032.2012.688060.

Endres, D.E., Cozen, B., Barnett, J.T., O'Byrne, M., and Peterson, T.R. (2016) Communicating Energy in a Climate (of) Crisis. *Annals of the International Communication Association* 40(1): 419–447. DOI: 10.1080/23808985.2015.11735267.

Goodwin, J. and Dahlstrom, M.F. (2014) Communication Strategies for Earning Trust in Climate Change Debates. *Wiley Interdisciplinary Reviews: Climate Change* 5(1): 151–160. DOI: 10.1002/wcc.262.

Hart, P.S. and Nisbet, E.C. (2012) Boomerang Effects in Science Communication: How Motivated Reasoning and Identity Cues Amplify Opinion Polarization about Climate Mitigation Policies. *Communication Research* 39(6): 701–723.

Hine, D.W., Reser, J.P., Morrison, M., Phillips, W.J., Nunn, P., and Cooksey, R. (2014) Audience Segmentation and Climate Change Communication: Conceptual and Methodological Considerations. *Wiley Interdisciplinary Reviews: Climate Change* 5(4): 441–459.

Hoffman, A.J. (2011) Talking Past Each Other? Cultural Framing of Skeptical and Convinced Logics in the Climate Change Debate. *Organization & Environment* 24(1): 3–33. DOI: 10.1177/1086026611404336.

Holden, E. (2018) Climate Change Skeptics Run the Trump Administration. *POLITICO*, March 7. Available at: http://politi.co/2oVqXYd (accessed November 20, 2018).

Johannesen, R.L. (1974) Attitude of Speaker toward Audience: A Significant Concept for Contemporary Rhetorical Theory and Criticism. *Communication Studies* 25(2): 95–104.

Julian [pseudonym]. (2018) personal communication [digital exchange].

Juris, J.S. (2008) *Networking Futures: The Movements against Corporate Globalization.* Durham, NC: Duke University Press.

Kirk, K. (2018) Focus on Those with an Open Mind. Yale Climate Connections, November 19. Available at: www.yaleclimateconnections.org/2018/11/focus-on-those-with-an-open-mind/ (accessed December 19, 2018).

Konisky, D.M. (2018) The Greening of Christianity? A Study of Environmental Attitudes over Time. *Environmental Politics* 27(2): 267–291.

Leiserowitz, A.A., Maibach, E., Roser-Renouf, C., Feinberg, G., and Rosenthal, S. (2016) *Global Warming's Six Americas.* New Haven, CT: Yale Project of Climate Change, Yale School of Forestry & Environmental Studies and the Center for Climate Change Communication, George Mason University. Available at: http://climatecommunication.yale.edu/visualizations-data/global-warmings-six-americas-november-2016/ (accessed January 28, 2019).

Leiserowitz, A.A., Maibach, E., Roser-Renouf, C., and Smith, N. (2011) *Global Warming's Six Americas.* New Haven, CT: Yale Project of Climate Change, Yale School of Forestry & Environmental Studies and the Center for Climate Change Communication, George Mason University. Available at: http://environment.yale.edu/climate/files/Six AmericasMay2011.pdf (accessed April 6, 2013).

McFadden. B.R. (2016) Examining the Gap between Science and Public Opinion about Genetically Modified Food and Global Warming. *PLOS ONE* 11(11): e0166140. DOI: 10.1371/journal.pone.0166140.

Meyer, M. (2012) Aristotle's Rhetoric. *Topoi* 31(2): 249–252.

Myers, T.A., Nisbet, M.C., Maibach, E.W., and Leiserowitz, A. (2012) A Public Health Frame Arouses Hopeful Emotions about Climate Change. *Climatic Change* 113(3): 1105–1112. DOI: 10.1007/s10584-012-0513-6.

Ratcliffe, K. (1999) Rhetorical Listening: A Trope for Interpretive Invention and a "Code of Cross-Cultural Conduct." *College Composition and Communication* 51(2): 195–224. DOI: 10.2307/359039.

Ross, D.G. (2013) Common Topics and Commonplaces of Environmental Rhetoric. *Written Communication* 30(1): 91–131. DOI: 10.1177/0741088312465376.

Sangalang, A. and Bloomfield, E.F. (2018) Mother Goose and Mother Nature: Designing Stories to Communicate Information about Climate Change. *Communication Studies* 69(5): 583–604.

Schwarze, S. (2006) Environmental Melodrama. *Quarterly Journal of Speech* 92(3): 239–261. DOI: 10.1080/00335630600938609.

Taylor, C.A. (1992) Of Audience, Expertise and Authority: The Evolving Creationism Debate. *Quarterly Journal of Speech* 78(3): 277–295. DOI: 10.1080/00335639209383997.

Wardekker, J.A., Petersen, A.C., and van der Sluijs, J.P. (2009) Ethics and Public Perception of Climate Change: Exploring the Christian Voices in the US Public Debate. *Global Environmental Change* 19(4): 512–521. DOI: 10.1016/j.gloenvcha.2009.07.008.

Executive summary

Talking points and strategies for engaging in climate change conversations

This chapter is meant to serve as a reference guide for the main strategies I propose for engaging religiously-motivated climate deniers, motivating religious environmentalists to act, and promoting effective climate communication across audiences. In what follows, I include brief summaries of the theoretical and experimental justification for each strategy, based on the longer explanations in the preceding chapters, organized by my categories of separators, bargainers, and harmonizers, and strategies that may work across categories. These strategies are aimed primarily at addressing climate skepticism and climate apathy in Christian communities, but I hope that others may apply them outside of religious groups and beyond climate contexts to further test their utility in starting, maintaining, and reaching understanding and respect in conversations. Because these strategies directly address how people respond to scientific information, it is my hope that these strategies may be successful outside of the environmental scope in which they have been proposed in this book, and that these strategies can prompt the rethinking of how we communicate science to the public. I also hope that the list will continue to evolve over time in both my future work and the work of others who take up these important questions of how to have more frequent and more productive climate conversations.

While these strategies are important starting points, each conversation, each newspaper article, each interview, and each instance of public and interpersonal communication will be unique and should take into consideration the particular audience, topic, and situation. Because rhetoric is highly situational and contextual, there will never be a panacea that instantly and completely removes skepticism, doubt, and denial from the public sphere. But, that does not mean we should abandon all hope of improving scientific communication or shut down opportunities for conversation. For scholars or practitioners of climate communication, this chapter can serve as a guide for teaching and sharing productive communication strategies. First, I outline general strategies for communicating about the climate more effectively that should be applicable to most conversational and public communication contexts. Then, I summarize the specific strategies proposed and used in previous chapters for addressing separators, bargainers, and harmonizers, respectively.

General strategies for climate communication

1 Treat conversations as dialogues instead of monologues. Approach interactions (even if they are one-way communicative events) with a mindset of openness, listening, and valuing the audience

Johannesen (1974) argued that a speaker's attitude toward the audience directly influences how persuasive messages are. If we treat communication as a mono-logue (e.g., thinking that only our opinions are important to discuss), what we communicate might come off as a patronizing lecture instead of an opportunity for engagement. In approaching a conversation as a dialogue, we should focus on the shared goal of understanding rather than persuasion. This requires that we not only speak, but listen and remain open to alternative perspectives (Ratcliffe 1999). Goodwin and Dahlstrom (2014) argue that we must give trust in order to receive trust when approaching climate communication. Trust can be fostered through people "committing themselves to an on-going relationship" with their dialogue partner (Goodwin and Dahlstrom 2014: 151). One tip for entering con-versations as a dialogue is to look beyond the audience's skepticism and focus on the person. We should also try to approach conversations as opportunities for learning – even if we are not open to changing our minds on the truth of climate change, we may learn new information about how people come to understand, process, and respond to climate change messages. If we see our dialogue part-ners or our larger audience as valuable, important, and worthy of our engage-ment, we open up opportunities for our messages to be clearly and genuinely heard. This strategy, of course, implies that we are not engaging in argument as a form of coercion or manipulation, but in argument as a process of reaching a mutually beneficial understanding (Brockriede 1972).

2 Learn about the audience to locate their underlying values and then connect those values to environmental issues

Perelman and Olbrechts-Tyteca (1957) argue that values are key for discovering how people make decisions. For example, if someone prioritizes family above all else, they are likely to turn to that value in decision-making instead of wealth, the environment, or other values. By learning about our audience and their likely values, we can tailor our messages to activate those values and link those values to environmental beliefs and behaviors. We should also think of values in terms of scales and hierarchies, where skeptics may likely report that they care about the environment but would be willing to sacrifice the environment if a higher-ranked value is at stake. One way to learn about the values of a particular person in an interpersonal conversation is simply to ask them what is important to them. If the audience is a larger group, demographic research (e.g., age, geographic location, political party, occupation) can help identify likely values and prior-ities. One can also always fall back on universal values such as family, money, life, and happiness. This strategy comes with a caveat, however, which is that linking values to environmental topics can backfire if the audience finds the appeals to be insincere or inauthentic. Myers *et al.* (2012: 1111) found that some people perceive such linking as an attempt "to co-opt values they care strongly about, thereby producing a negative reaction." Therefore, when making such links, it is important to establish authenticity and mutual understanding.

3 Find points of common identity so people can feel connected to the conversation and the topic and be motivated to act in response

Many studies have supported the idea that finding connections between people help them "identify" with one another, which will then increase a person's interest and openness to persuasion (Burke 1969). For example, Hart and Nisbet (2012) found that linking readers to impacts of climate change in communities similar to theirs was more persuasive than asking them to think about markedly different populations, to the extent that even political affiliation's influence was dampened. A similar study found that narrowing organ donation statistics from the national level to the participants' state increased the persuasiveness of a message to sign up for organ donation (Dillow and Weber 2016). Similar to the caveat above, finding points of common identity may also come off as insincere, especially when audiences perceive a certain identity as more accurately belonging to a different political affiliation (Tormala *et al.* 2006). Thus, it is important to establish open and welcoming conversation to deflect potential misreadings of invoking values and identities.

Strategies for addressing separators

1 Ask questions to learn more about the person you are engaging. People want to see themselves as valued conversation partners and not as targets to be convinced or persuaded

Questions serve two purposes. First, they provide the opportunity to learn more about the person to make use of that information for the following two strategies. Second, they help break down walls and avoid boomerang effects (Hart and Nisbet 2012) that may end conversations before they begin. In line with our general strategy of dialogue, we do not want conversations to turn into "question and answer" sessions or formal interviews. There should be healthy back-and-forth where people are open to answering questions about themselves and extending questions to us. In a public setting, rhetorical questions that prompt separators to think about their values, identities, and positions on issues can help bring certain premises to the surface for the next strategy.

2 Offer premises (parts of arguments) that align with existing beliefs

This strategy makes use of the persuasive power of enthymemes, which involve the audience in their own persuasion by not fully explaining the argument. By leaving a part of the argument out, audiences can fill in their own answers that res-onate with them. Long (1983) argues that audience-centered persuasion involves the audience in its own convincing. Enthymemes have been long-studied as highly persuasive rhetorical devices (Bloomfield and Sangalang 2014). Instead of com-pleting the argument, we can ask questions and repeat back to our dialogue part-ners their own beliefs that can lead to environmentally-oriented conclusions. For example, a person who professes that they value human life over the environment can be prompted to consider to what extent environmental degradation is detri-mental to the wellbeing of human life. In a public setting, we can offer likely pre-mises that would be adhered to by our audience through our discovery of common values and identities. For example, if speaking to a group of small business owners, one might appeal to the economy, job expansion, and opportunities for increased efficiency when businesses adopt eco-friendly practices. The audience can then fill in the rest of the argument by reflecting on how their business might similarly become more successful if adopting those same practices.

3 Make it personal by finding local, communal, and situational nuances related to your audience

While this is an overall strategy that should be used in climate communication in general, this strategy is particularly important for separators who believe that they are the gatekeepers of authority and truth. This strategy enables separators to feel personally involved in the topic. In addition to thinking about finding common values, it is also important to locate what Wang *et al.* (2018: 26) call "objects of care" that can evoke emotional connections between an individual and the topic of the environment. In order to protect these objects (e.g., auto-nomy, personal property, community institutions, landmarks, etc.), people open themselves up to hearing from a variety of perspectives focused on protecting that object of care. For example, climate skeptics who are conservative are likely to respond to concrete values rather than abstract ones (Bloomfield and Katula 2012), which makes locating specific, personalized objects potentially more con-vincing or engaging. Similar to the general strategy of finding points of "common identity," the strategy of "making it personal" lowers social distance between the speaker and audience and puts the audience in a frame of mind where they may be personally implicated in the topic (Dillow and Weber 2016).

Strategies for addressing bargainers

1 Work within frames by embracing competing narratives as opportunities to find overlap and convergence. If someone shifts an environmental conversation to economics, religion, or politics, we can modify our communication to engage those frames

People do not think, act, or communicate in a vacuum. As much as we would like to simplify the environment and tackle it as a single topic, climate change is truly a "wicked problem" that engages various disciplines, frames, and knowledge sources (Cagle and Tillery 2015). Recognizing the multiplicity of ways that climate skeptics make sense of climate change can prompt us to meet them where they are at. We cannot assume that our audience is uninformed or lacks environmental knowledge. Bargainers are likely to substitute climate authorities with economic authorities that support their frames of religious autonomy and the ethic of capitalism. Elliott (2014) argues for the utility of indirect arguments, which support environmental claims through economic, political, social, and cultural grounds. Activating those values may encourage bargainers to see their dominant frame as compatible with environmental thinking, because the environment is not presented as endangering their values and priorities or requiring sacrifice. This strategy engages the broader benefits of environmental protection, such as technological advancements, increased business opportunities, and innovation, without focusing on protecting the environment as a primary goal. While I agree that there are issues with constantly devaluing and overshadowing the importance of the environment in these discussions, this strategy of deflection and indirect persuasion can be a way to temper reluctance and rejection within certain skeptical communities.

2 Joining the revolution by recognizing the variability and flexibility of scientific knowledge as a way to guard against bargainers' refutation of scientific consensus

Bargainers tend to see the relationship between mainstream climate science and their perspectives of it through the lens of revolution, where their minority viewpoint will eventually become the dominant perspective. One way to engage bargainers is to embrace that revolutions happen and are valuable to driving scientific knowledge forward, but the current state of climate skepticism does not merit consideration as an imminent revolution. For example, we can uncouple the notion that the mere presence of skeptics, especially the minimal amount of climate skeptics, means that revolution will happen by pointing to other minority viewpoints such as the anti-vaccine movement or Flat Earthers. Simply because there are those that challenge mainstream scientific interpretations does not always mean current mainstream conclusions will be overthrown. Ceccarelli (2011: 213) argued that one strategy to refute those who wish to keep unproductive conversations going is to "acknowledge that debate is important to science" while also emphasizing that debates over climate change have already and extensively occurred and have "been decided against the dissenters."

3 Employ examples by finding concrete statistics, narratives, and similar situations where people have found success through environmentally-restorative actions, beliefs, and behaviors

People who oppose climate change due to economic or political reasons could be shown examples of how being environmentally-friendly has led to economic gains (Eccles *et al.* 2012) or how the environment does not have to be a political issue but a human issue, respectively. By breaking previously held "illusory correlations" that say that environmental science and an alternative narrative are incompatible, we can rewrite the story and provide new correlations through counter-examples (McFadden 2016). Providing concrete examples of productive political, economic, and religious intersections with the environment can help our dialogue partners or our larger audience start making those connections themselves.

Strategies for addressing harmonizers

1 Shift frames from private to public activism and encourage the audience to see private actions as part of larger public efforts

Stern (2000) argued that there are two types of environmental advocacy: (1) private-sphere activism consists of individual activities that are often kept in the home, and (2) public environmentalism consists of the public performance of an environmental identity that involves community- and societal-level engagement. Although harmonizers often agree that climate change is important to mitigate, they may direct their attention to private instead of public activism. By encouraging harmonizers to expand the scope of the private realm to include their family, friends, and church communities as extensions, we can encourage more prominent exposure and discussion of environmental values in trusted peer networks. Although harmonizers perceive some risk in "coming out" as environmentalist to their communities, we can try to assuage these fears by offering advice on how to bring up environmental topics and encourage them to start small with trusted interpersonal relationships. Although we do not want to undermine the value of individual actions, we also want to empower harmonizers to see that relatively simple and private actions such as having a conversation can have more profound effects than an individual alone can.

2 Highlight the urgency of action and the agency of people to contribute to the mitigation of climate change

Harmonizers already believe the environment is important but may not feel that they can take meaningful action or that climate change is a pressing concern that requires immediate action. Harmonizers are open to hearing statistics and scientific evidence to convince them of climate changes urgency because they value scientific information. They can also be provided Bible verses and exemplar religious leaders about the urgency of the issue. Using a blend of scientific and religious justification can motivate harmonizers to think of the issue of climate change as pressing and requiring immediate action. Urgency can also be highlighted by providing examples of the damages that have already been done to people around the world and evoking loss frames of what may be lost if immediate action is not taken. We can also tell harmonizers about the threats that Christian climate skeptics pose to highlight the need for urgent action to counter them. Because people in developed countries may not have experienced consequences from climate change (or may not recognize them as such), we can employ examples of those already suffering to ignite the urgency to act.

3 Engage harmonizers' ecological beliefs by emphasizing global concerns

Providing examples from around the world can not only prompt urgent action, but can draw on harmonizers' ecological beliefs that all life is connected. Waddell (1990) argued that appeals to emotion lead to conviction toward actions. We can activate harmonizers' emotional involvement by providing examples of communities around the world already affected by climate change and can link the environment to existing religious values such as charity, social justice, and abortion issues (Thomas 2014). Elevating the problem of climate change to a global issue may also promote considering the solution on an international level, prompting a reconsideration of the efficacy of turning solely to personal behavior changes.

Conclusion

To end this book with a final thought, it is my hope that we can embrace a comic approach toward climate skeptics where we consider them mistaken instead of sinful (Burke 1984; Carlson 1986; Desilet and Appel 2011). It is challenging to come to terms with the various reasons why people may reject or deny the overwhelming truth and severity of climate change. And yet, some do. If we change the frame of our approach and the goal of our interactions, we might find a way forward that embraces the person behind the denial and offers hope for productive communication. Putting in the effort, being vulnerable, seeking dialogue, and simply listening may help engage a portion of the population that often feel silenced, shut out, and unacknowledged. To change our frame, we must enter conversations free from bias or preconceived expectations of our dialogue partners' beliefs, values, or positions. We must also listen, not just for a talking point to jump in on, but to understand the perspective they are coming from and what values or identities they feel are threatened by environmentalism and mainstream conceptions of climate change. If we can see the person behind the opinion, we might locate mutual understanding and dialogue, where the separators' melodramatic war and the bargainers' persistent hope for revolution is replaced with a sense of harmony, but one that calls for immediate, public action.

References

Bloomfield, E.F. and Katula, R.A. (2012) Rhetorical Criticism of the 2008 Presidential Campaign: Establishing Premises of Agreement in Announcement Speeches. *Communication Research Reports* 29(2): 139–147. DOI: 10.1080/08824096.2012.667775.

Bloomfield, E.F. and Sangalang, A. (2014) Juxtaposition as Visual Argument: Health Rhetoric in Super Size Me and Fat Head. *Argumentation and Advocacy* 50(3): 141–156.

Brockriede, W. (1972) Arguers as Lovers. *Philosophy & Rhetoric* 5(1): 1–11.

Burke, K. (1969) *A Rhetoric of Motives. 1950.* Berkeley, CA: University of California Press.

Burke, K. (1984) *Attitudes toward History.* Berkeley, CA: University of California Press.

Cagle, L.E. and Tillery, D. (2015) Climate Change Research across Disciplines: The Value and Uses of Multidisciplinary Research Reviews for Technical Communication. *Technical Communication Quarterly* 24(2): 147–163. DOI: 10.1080/10572252. 2015.1001296.

Carlson, A.C. (1986) Gandhi and the Comic Frame: "Ad Bellum Purificandum." *Quarterly Journal of Speech* 72(4): 446–455. DOI: 10.1080/00335638609383787.

Ceccarelli, L. (2011) Manufactured Scientific Controversy: Science, Rhetoric, and Public Debate. *Rhetoric Public Affairs* 14(2): 195–228.

Desilet, G. and Appel, E.C. (2011) Choosing a Rhetoric of the Enemy: Kenneth Burke's Comic Frame, Warrantable Outrage, and the Problem of Scapegoating. *Rhetoric Society Quarterly* 41(4): 340–362. DOI: 10.1080/02773945.2011.596177.

Dillow, M.R. and Weber, K. (2016) An Experimental Investigation of Social Identification on College Student Organ Donor Decisions. *Communication Research Reports* 33(3): 239–246. DOI: 10.1080/08824096.2016.1186630.

Eccles, R.G., Ioannou, I., and Serafeim, G. (2012) *The Impact of a Corporate Culture of Sustainability on Corporate Behavior and Performance.* Cambridge, MA: National Bureau of Economic Research.

Elliott, K.C. (2014) Anthropocentric Indirect Arguments for Environmental Protection. *Ethics, Policy & Environment* 17(3): 243–260. DOI: 10.1080/21550085.2014.955311.

Goodwin, J. and Dahlstrom, M.F. (2014) Communication Strategies for Earning Trust in Climate Change Debates. *Wiley Interdisciplinary Reviews: Climate Change* 5(1): 151–160. DOI: 10.1002/wcc.262.

Hart, P.S. and Nisbet, E.C. (2012) Boomerang Effects in Science Communication: How Motivated Reasoning and Identity Cues Amplify Opinion Polarization about Climate Mitigation Policies. *Communication Research* 39(6): 701–723.

Johannesen, R.L. (1974) Attitude of Speaker toward Audience: A Significant Concept for Contemporary Rhetorical Theory and Criticism. *Communication Studies* 25(2): 95–104.

Long, R. (1983) The Role of Audience in Chaim Perelman's New Rhetoric. *Journal of Advanced Composition* 4: 107–117.

McFadden, B.R. (2016) Examining the Gap between Science and Public Opinion about Genetically Modified Food and Global Warming. *PLOS ONE* 11(11): e0166140. DOI: 10.1371/journal.pone.0166140.

Myers, T.A., Nisbet, M.C., Maibach, E.W., and Leiserowitz, A. (2012) A Public Health Frame Arouses Hopeful Emotions about Climate Change. *Climatic Change* 113(3): 1105–1112. DOI: 10.1007/s10584-012-0513-6.

Perelman, C. and Olbrechts-Tyteca, L. (1957) The New Rhetoric. *Philosophy Today* 2(1): 4–10. DOI: 10.5840/philtoday195711/42.

Ratcliffe, K. (1999) Rhetorical Listening: A Trope for Interpretive Invention and a "Code of Cross-Cultural Conduct." *College Composition and Communication* 51(2): 195–224. DOI: 10.2307/359039.

Stern, P.C. (2000) New Environmental Theories: Toward a Coherent Theory of Environmentally Significant Behavior. *Journal of Social Issues* 56(3): 407–424. DOI: 10.1111/0022-4537.00175.

Thomas, M. (2014) Pro-Life Equals Pro-Planet for This Green Evangelical Leader. *Grist*, December 26. Available at: https://grist.org/living/pro-life-equals-pro-planet-for-this-green-evangelical-leader/ (accessed March 16, 2018).

Tormala, Z.L., Briñol, P., and Petty, R.E. (2006) When Credibility Attacks: The Reverse Impact of Source Credibility on Persuasion. *Journal of Experimental Social Psychology* 42(5): 684–691. DOI: 10.1016/j.jesp.2005.10.005.

Waddell, C. (1990) The Role of Pathos in the Decision-Making Process: A Study in the Rhetoric of Science Policy. *Quarterly Journal of Speech* 76(4): 381–400. DOI: 10.1080/00335639009383932.

Wang, S., Leviston, Z., Hurlstone, M., Lawrence, C., and Walker, I. (2018) Emotions Predict Policy Support: Why It Matters How People Feel about Climate Change. *Global Environmental Change* 50: 25–40. DOI: 10.1016/j.gloenvcha.2018.03.002.

Index